T0178119

Springer Tracts in Civil Engineering

Springer Tracts in Civil Engineering (STCE) publishes the latest developments in Civil Engineering - quickly, informally and in top quality. The series scope includes monographs, professional books, graduate textbooks and edited volumes, as well as outstanding PhD theses. Its goal is to cover all the main branches of civil engineering, both theoretical and applied, including:

- Construction and Structural Mechanics
- Building Materials
- Concrete, Steel and Timber Structures
- Geotechnical Engineering
- Earthquake Engineering
- Coastal Engineering; Ocean and Offshore Engineering
- Hydraulics, Hydrology and Water Resources Engineering
- Environmental Engineering and Sustainability
- Structural Health and Monitoring
- Surveying and Geographical Information Systems
- Heating, Ventilation and Air Conditioning (HVAC)
- Transportation and Traffic
- Risk Analysis
- Safety and Security

Indexed by Scopus

To submit a proposal or request further information, please contact: Pierpaolo Riva at Pierpaolo.Riva@springer.com, or Li Shen at Li.Shen@springer.com

More information about this series at http://www.springer.com/series/15088

Bruno Daniotti · Alberto Pavan ·
Sonia Lupica Spagnolo ·
Vittorio Caffi · Daniela Pasini ·
Claudio Mirarchi

BIM-Based Collaborative Building Process Management

Bruno Daniotti
Dipartimento ABC
Politecnico di Milano
Milan, Italy

Sonia Lupica Spagnolo
Dipartimento ABC
Politecnico di Milano
Milan, Italy

Daniela Pasini
Dipartimento ABC
Politecnico di Milano
Milan, Italy

Alberto Pavan
Dipartimento ABC
Politecnico di Milano
Milan, Italy

Vittorio Caffi ⓘD
Dipartimento ABC
Politecnico di Milano
Milan, Italy

Claudio Mirarchi
Dipartimento ABC
Politecnico di Milano
Milan, Italy

ISSN 2366-259X   ISSN 2366-2603   (electronic)
Springer Tracts in Civil Engineering
ISBN 978-3-030-32891-7   ISBN 978-3-030-32889-4   (eBook)
https://doi.org/10.1007/978-3-030-32889-4

This Springer imprint is published by the registered company Springer Nature Switzerland AG
The registered company address is: Gewerbestrasse 11, 6330 Cham, Switzerland

Introduction

It is now 45 years (1974–2019) since C. Eastman theorized what would later become known in the world as BIM (Building Information Modelling). After 45 years, it seems strange that a methodology or the use of some related tools are still seen as an exception to practice, as an eternal novelty or a hope, partly (evidently) unfinished.

Alongside to some excellences as USA, UK and Scandinavian countries, but nowadays also France and Italy even if still little known, there are still large areas in the world where BIM is not applied with continuity. In the same way, we should consider that also in the countries that are most advanced in this change, it is not uniform but spreads like a cheetah stain. First, it is integrated in the larger design firms and projects and then, with difficulties, it starts its diffusion to the entire market and all the other small stakeholders. Starting from architects and engineers, followed by the construction companies and only at the end the clients.

On the one hand, it will be said that BIM is a silent revolution, and thus, it needs more time to emerge. On the other hand, that there was a push to accelerate too much a sector (the construction sector) that was not ready for a so radical change and thus we should give the necessary time for a double jump; the one related to the mentality and that related to the tools.

Anyway, in both visions, the required time appears incompatible if compared to the timing of digitalization. Forty-five years compared to 10 or 5 years of every other sector; think for example to the mobile phone sector.

The fourth industrial revolution (Industry 4.0), the revolution of data and connection, has definitely revealed the critical issues of the construction sector and today it cannot hope to be so different to the other production or services sectors.

Even the construction "industry" must fully enter the diffused digitization. If not for its own needs, at least because the other sectors are requiring this change and they need that also the last bulwark of the tradition falls entering in the new paradigm.

In the same way, the tools in the Architecture, Engineering and Construction (AEC) sector must interact and enter in the web to generate a critical mass of data. The buildings must communicate the performance of the component products (air conditioning, lighting, etc.) to the production, to be monitored and improved, as well as the habits of the users (use of meeting rooms, spaces that are over or under used, etc.) to improve the management and anticipate the needs, etc.

The new AEC software is nowadays all "BIM", or they tend to a plus or minus (or effective) "BIMization" of their product. The big projects at world level are developed in BIM. BIM is used by big design firms and the largest construction companies. There is a large spectrum of standards and legislation about BIM. The new big public works are commissioned in BIM.

There isn't a questionnaire, report or analysis, from which the inevitability of BIM is not shown for all type of product (infrastructures, buildings, etc.) or work (new, renovation, etc.) and for any type of subject (customers, designers, construction companies, etc.).

BIM is at the same time a mature reality, by commercial, and an incomplete revolution, in its entirety.

In actual fact, without considering the nowadays not rare excellences that obviously exist also in the BIM area, on the one hand, there is still an often uncertain use of the tools, very specialist, based on approximations by trial and error, a diffuse manual and practical dexterity and a limited knowledge of the complexity in the informatic (and informative) terms that is behind the software front-end.

On the other hand, still exists an issue in the approach to the change. This last is often linked to the voluntary enthusiasm, that risks to be inefficient due to the lack of diffusion and of interaction of the system, or linked to the legislative imposition that on the other hand, risk the typical inefficiency of the top-down obligations not participated by the bottom.

In this, the university can play a fundamental role in the diffusion of the knowledge focused on the use of the tools and their applications in the method, standing above the needs of the moment. To avoid the marginalization due to contingent operational problems, the university should focus on solving the current issues while looking in a long term perspective paving the way for the future evolutions.

A mixed approach of practice and theory, Information Technology (IT) and AEC, that must be maintained linked to permanently involve the sector. Otherwise, the sector could become a disappointed "not BIM user" as well as it could become an enthusiastic "unaware user" of the tools. Both these perspectives must be avoided, because pointless for the sector that could be preyed by the other ones.

Google, Amazon and Alibaba could, in a not so far future, remove the intermediation between the manufacturing industry and the production of components towards the final users through the Artificial Intelligence (AI) for the design and acting as a global general contractor.

Too many sectors have been caught unprepared by the change, thinking it was impossible for them to be swept away, deluded by a presumed diversity (which later turned out to be ephemeral). Think of the hotel industry, with the thousands of rooms in the world of the "diffused" hotel created by Airbnb, as well as the transport, with the infinite non-professional drivers connected by Uber. The phone industry absorbed photography, Internet the retail, etc.

If the construction sector, as we know it today, will not be able to innovate and also to govern the digital change, others will do it for him. At least because it represents one of the primary needs of the humankind, the shelter, the "house" in a broad sense, and because it is a sector that generates an added value with a market value between 10 and 13% of the world GDP.[1]

If the first papers and the first standards were too tied to the novelty on the informatics and software tools, operativity of the IT world, today perhaps we are witnessing an opposite excess in the theorization of the method, towards its academic formalization in the AEC sense, which leaves the sector lacking in an informatics feedback and application opportunities for the operators.

This demonstrates a still strong uncertainty of the sector in the direction to be taken.

Think about the need to share data, information and consequently knowledge. This is a common problem to all disciplines and sectors and thus obviously for the AEC one. Above all in the AEC sector, where there has always been a strong issue in the interaction between subjects, products and services. The legend of the Babel tower in the Genesis is maybe the most striking and enlightening case.

In the BIM world, this critical issue (informative) is "solved", in conceptual and standard terms, with the introduction of a digital environment devoted to share, defined as Common Data Environment (CDE) first in the BS 1192-1:2007 and then in the PAS 1192-2:2013.

Since then, though, exist the CDEs described by the standard, rules, the ones described by the academia, method, and the ones developed by the software houses, tool. Now, these CDEs cannot remain similar but not equal for too long if the diversity is not in a different perspective in the solution of the problem but in the answer to the problem itself; as if they were three different problems.

In the same way, it is not feasible that the interaction between tools is still based on an intensive work of the human operator due to the inefficiency of the not graphic software in the construction domain. The lack of an AEC ontology cannot be solved using only the IFC classes and consequently blaming at the latter being ineffective due to objectives that by its nature, it cannot have or resolve.

The transition to the object-oriented programming, with the specific AEC objects (walls, roofs, doors, windows, etc.) and not geometric (or related to the geometric representation with lines, surfaces, volumes), created by the world of graphical software, or of BIM authoring software,[2] has not been really completed, for example by the software specialized in the scheduling of the work and in the evaluation of its costs.

Thus, it is difficult to say to the entire set of stakeholders that BIM is not only 3D, but also 4D (time) and 5D (costs), if not simply referring to the standards and to the method. Because, the potential of the BIM authoring tools, to control and

---

[1]Gross Domestic Product.

[2]Revit, Archicad, Allplan, Miscrostation, Cathia, Edificius, etc.

manage the objects, are nowadays not available in the cost estimating and project management software.

When will we be able to work with a Project Management (PM) tool where the objects are the activities of the AEC world (construction, demolition, laying, painting, etc.) and thus able to comprehend with a sort of autonomy the possible interactions and critical points in the connection between physical objects defined in BIM authoring tools (doors, windows, walls, slabs, etc.) and the activities in the PM one?

If there are still areas of distrust or inefficiency in the use of BIM maybe the BIM world should ask to itself new questions with reference to the ones that for years have been circulating: what is it, who use it, why he or she use it, what advantages, etc.?

Have we solved all the real and applicative issues of the sector?

Or maybe are we still focused on those related to the design, forgetting, in the practice, construction and operation?

If we look to the number and to the full functionality of the software available in the market maybe, we are only telling (rules and methods) that BIM is made with a focus to the management of the construction works and not only for the design stage.

When BIM is used as graphical support for the management, in support for the data representation, and it is not integrated with the management system where the data are managed, stored, used, etc. (with some exceptions that fortunately exist) we are still in an intermediate informatic passage. With the BIM viewed as a support to the facility management software or to the real estate one, substituting the old 2D representations (pdf or dwg), but still without the perception of a software focused on strategy and operation (asset, property, facility) in the AEC world, native BIM.

It is like the difference in the manual indexing of Yahoo and the advent of the Google algorithms. Not that Yahoo was not "Internet" but an Internet still reasoned as if it were a "big" database of information (not "infinite").

Moreover, what is the margin of auto-resolution of issues we are still asking to the average operator (not the most evolved one with big capacity in the investments)?

Each not codified solution (generic objects, proprietary objects, etc.) surely creates a competitive advantage and a knowledge advantage for the single user, optimal for his or her market. However, this is clearly in contrast with a world that bases its nature on the exchange of data and the collaboration of the subjects. In this case it feels not (or not only) the lack of an open informatic language (actually well covered by the IFC for some aspects, at least for the geometric ones) but the lack of an ontology for the sector that can be used to derive common objects of every type, also not of BIM Authoring. In this direction, several young researchers (like J. Beets, P. Pouwels, L. van Berlo, G. F. Schneider, M. H. Rasmussen, E. Petrova, C. Mirarchi, etc.) are making good strides, but the market and the software houses are perhaps not cooperating enough.

Proprietary languages, rules and guidelines, as well as proprietary libraries, are the death of BIM if not united by a common "dictionary". Ontology and common classes, a sharing open language, then free competition for the generation and integration of solutions of more performing objects.

If BIM still does not represent the "normality" in the AEC sector, it cannot be always the fault of the stakeholders and of the backwardness of the sector. On the other hand, also other sectors, not least the agriculture, have long addressed the widespread digitization without a particular education level of the employees or the introduction of technological revolutions in the final product to which it has been applied (generic mechanical components).

A part (not a small one) of the guilt in the slow diffusion of BIM can be also assigned to those that are dealing with BIM, due to the inability in dealing and solving many of the aspects related to its use in the day-to-day work and in its integration and impact on the other company functions beyond design: whether operational or managerial.

This book presents the consolidated state of the art of BIM. This is useful in the definition of a knowledge base of the common subject and deals with many of the critical issues above mentioned providing a possible solution and trying to take an extra step required to consolidate the introduction of the digitization in the AEC sector.

The book is the result of common work in all its aspects.

In detail, has been helpful the specific disciplinary contribution of the authors as following listed:

- Introduction, Bruno Daniotti, Claudio Mirarchi;
- Evolution of the building sector due to digitalization, Alberto Pavan, Bruno Daniotti;
- Standardized structure for collecting information according to specific BIM uses (Technical datasheet), Sonia Lupica Spagnolo;
- Standardized guidelines for the creation of BIM objects, Vittorio Caffi;
- Collaborative working in a BIM environment (BIM platform), Alberto Pavan, Claudio Mirarchi;
- Benefits and challenges in implementing BIM in Design, Alberto Pavan;
- Benefits and challenges of BIM in construction, Claudio Mirarchi;
- Benefits and challenges using BIM for operation and maintenance, Daniela Pasini.

Contents

Chapter 1
Evolution of the Building Sector Due to Digitalization

Abstract Today, perhaps more than ever before, a technological revolution can modify the building sector in all its single aspects, greatly affecting services, production, supplies. Freehand drawing, drafting machines or CAD have represented innovative tools in graphic representations. In such cases, the evolution of tools for the productivity of the sector has improved and quickened the actual design, but nothing more. With BIM, instead, the innovation of tools has entailed a methodological innovation for the whole sector, owing to virtual reality simulations, and not only to graphic representations. Several basic principles and their various evolutions worldwide can help understand the BIM phenomenon in order to achieve its mature use.

1.1 Introduction to BIM

As all technological revolutions, also the massive introduction of the Building Information Modeling—BIM [15] in the building sector is the result of a series of concomitant events that unblocked its quantitative dissemination (consistent use) and especially its qualitative dissemination (coherent use).

With regard to the technological aspect, if the introduction of CAD [10] in the "design" environment dates back to 1972 with Sutherland's "Sketchpad" (MIT, Massachusetts Institute Technology), the theorization of BIM [4] is not as recent as one might think, since it dates back to 1974 (Fig. 1.1), thus prior to the creation of AutoCAD (1982, Autodesk), the most known CAD software used in the AEC sector.

In the following years, other software houses began to develop what today is known as BIM, starting from Graphisoft with Archicad (1987), that is the evolution of its Radar CH CAD. This was followed by Nemetschetck's Open BIM (1997) and Bentley's Triform BIM (1998).

In terms of commercial dissemination, though, the true turning point took place when also Autodesk decided to approach BIM when it purchased Revit Technology Corporation and its "REVIT" software (2002). Autodesk's purchasing policy led it, on the one hand, to its current conformation as multiple tool platform of integrated

© Springer Nature Switzerland AG 2020
B. Daniotti et al., *BIM-Based Collaborative Building Process Management*, Springer Tracts in Civil Engineering,
https://doi.org/10.1007/978-3-030-32889-4_1

1962	Sketchpad – CAD (MIT)	Ivan Sutherland
1974	An Outline of the Building Description System - BDS	Charles Eastman
1982	AutoCAD/ USA / Autodesk	John Walker
1984	Radar CH (1987 BIM ARCHICAD) / Ungary / Graphisoft	Gábor Bojár
1984	ALLPLAN – (1997 BIM O.P.E.N.) / Germany / Nemetschetck	Georg Nemetschetck
1985	Microstation 1.0 (1998 BIM TRIFORM) / G.B. / Bentley	Keith A. Bentley
1997	REVIT Technology / USA / C.River Soft. (2002 Autodesk)	Irwin Jungreis, Leonid Raiz
2002	Building Information Modeling - BIM (1992 Nederveen-Tolman)	Jerry Laiserin

Fig. 1.1 CAD to BIM evolution

solutions; on the other hand, it led it to a long maturation that required years due to the time necessary for the mentioned aggregation of the many software originated autonomously to take place. Moreover, the following 10 years were characterized by a period of substantial inertia with regard to the global dissemination of the BIM phenomenon broadly speaking, perhaps with the sole exclusion of the USA market (which still is the most important worldwide) and niches such as the Scandinavian countries (with a fast evolution but, as to numbers, with a derisory commercial incidence).

This could lead to think that the dissemination of BIM is due, more than anything, to the marketing needs of software houses in the AEC sector, although consistent with an actual and important technological evolution. It is somewhat similar to what happened when Microsoft passed from 32 to 64 bit, o from Windows'98 to the current versions, such as "10." Indeed, such technological changes require a market diversification.

Actually, with the dissemination of web and telephone devices among users and now with Industry 4.0 in the production sector, what makes and will always make the difference is the interest that the ICT field (Information Communication Technology) has towards the building and real estate sector.

In an increasingly digitalized world, with a power relationship always more listed towards new technologies and communication, the presence of a sector still not much digitally developed represents a very vast and strongly attractive market for the main IT (and media) companies worldwide.

Google researches—or those of its satellite spin-offs, such as FLux, Tango, or Project Sun Roof, up to Google Glasses for the AEC sector in the more professional environment up to the various home assistance systems such as Google Home, Amazon Echo addressed to final users—are showing an increasing interest for the building sector, homes, offices. Indeed, they are searching for a stronger presence in a market that is worth about 13.5% of the world GDP, and that is strongly developing in emerging countries and economies (the USA and China ranking first, Africa in the upcoming future).

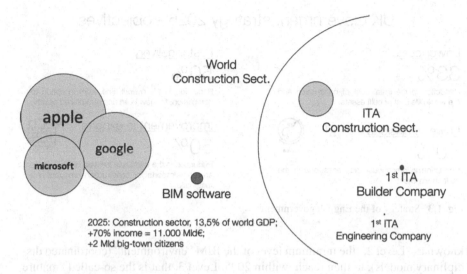

World
Construction Sect.

apple

google

microsoft

BIM software

ITA
Construction Sect.

1st ITA
Builder Company

2025: Construction sector, 13,5% of world GDP;
+70% income = 11.000 Mld€;
+2 Mld big-town citizens

1st ITA
Engineering Company

Fig. 1.2 Building sector and ICT sector in comparison per capitalization

Therefore, the change is not being imposed by the software houses in the AEC sector. Of course, though, they are spurring the direct digitalization of the sector with an increasing widespread interaction with the other markets, services and productions (Fig. 1.2).

Over the years, alongside software developments, a system of voluntary regulations has been structured, starting from the international ISO regulations. Also a series of mandatory regulations have been drawn up, starting from the EU Directive of 2014 [3] on public procurement. If, on the one hand, such regulations support the technical part, on the other hand they support and set the timing of the systematic introduction of BIM in the public and private sectors.

BIM was already supported in the public sector in Finland, Norway and Denmark, but the decisive turning point took place starting in 2011 with the UK Government's "Construction Strategy" and the objective to reach "*a deep change in the relationship between the public administration and the building industry*" in order to achieve a "*structural reduction of costs equal to at least 20%, starting from year 2016.*"

Therefore, in 2013 the UK Government communicated the strategy that it intended to adopt in order to relaunch the national building sector: starting from 2016, its intention was to pass on to BIM beginning from governmental public procurements above 5 million pounds, to then reach specific chain objectives within 2025, that is a 33% reduction in costs, a 50% reduction in overall timing, a 50% reduction in emissions, and a 50% increase in exportations (Fig. 1.3).

Following the drawing up of strategic guidelines, a specially devoted BIM Task Group was formed (Mark Bew, Adam Matthew, and others) [8], with the task to establish the timing and modalities for leading the whole sector and the Public Administration to transit from the CAD environment (defined as Level 0) to what is by now

UK Government strategy 2025 - objectives

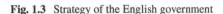

Lower costs

33%

Reduction in the initial cost of contruction and
the whole life cost of built assets.

Faster delivery

50%

Reduction in the overall time, from inception to
completion, for new build and refurbished assets.

Lower emissions

50%

Reduction in greenhouse gas emissions in the
built environment.

Improvement in exports

50%

Reduction in the trade gap between total Exports
and total Imports for construction products and
materials.

Fig. 1.3 Strategy of the English government

known as "Level 2," the minimum level of the BIM "environment" (coordinated disciplinary models), to then reach, within 2025, Level 3, that is the so-called "mature BIM."

The UK's acceleration was followed by France that, under the spur of the professionals of the sector, within the plan to relaunch the building sector (Plan de transition numerique dans le batimant), set the aim in 2014 to build 500 thousand new houses and restore another 500 thousand existing houses through BIM digitalization, appointing at the same time CSTB with the mission to digitalize the building sector (Mission Numérique et Bâtiment).

Following Great Britain and France, Germany's central Government formalized the forming of a specially devoted commission (Great Works Reform Commission) which not only has the task to implement the EU Directive on BIM, but it also has to carry out an overall digital reform of public procurements (Bauen Digital GmbH and Platform Digitales Bauen), in line with the country's industrial digitalization program (Industrie 4.0).

Under the UK's spur (2013), the EU Directive on Public Procurement of 2014 introduced this new industrial policy for the building sector also at community level; in fact, Art. 22, paragraph 4, states: "*For public works contracts and design contracts, Member States, may require the use of specific electronics tools, such as of* **building information** electronic **modeling tools** *or similar.*" The use is optional, but the explicit reference—although with slightly cryptic terms—allowed every public contracting authority, from that moment on, to promote calls for competition relating to BIM without the possibility to appeal to the European Court for the violation of free competition (reference to a specific technology not accessible to all competing subjects: BIM).

Consequently, in these years, in compliance with the EU directive, all Member States have been introducing BIM in their national Public Procurement Codes. In Italy's case, reference is made to Lgs.D. no. 50/2016, Art. 23, paragraph 1, letter h), "*use of* **specific electronic methods and tools, such as the modeling ones, in the building sector and infrastructure**," followed by a specific Decree (Ministerial

Table 1.1 Agenda for the mandatory introduction of BIM in public procurements

Year	Typology	Amount
2019	Complex works	≥ 100 million €
2020	Complex works	≥ 50 million €
2021	Complex works	≥ 15 million €
2022	Works	$\geq$ threshold Art.35 Lgs.D. 50/2016
2023	Works	≥ 1 million €
2025	Works	< 1 million €

Decree no. 560/2017: Baratono) on the gradual obligation to insert BIM in public procurements according to an established agenda (Table 1.1).

Always at European level, it is important to mention the "**Handbook for the introduction of BIM by the European Public Sector**" [5], drawn up by EU-BIM Task Group, the specially formed Commission composed of the major public contracting authorities. The Task Group was requested by Great Britain with the aim to understand how to launch common strategies at community level on great works: first of all railways and roads, since they are the means of connection and union between Member States. Indeed, such union is not only physical but also relating to technologies, building systems, design, procedures, etc. It is the ideal environment for an integration and interoperability environment such as BIM.

With regard to the voluntary technical regulations, there is a consolidated framework of regulations of reference, drawn up at international tables—ISO (TC59/SC13/BIM)—and community tables—CEN (TC442/BIM), where Italy participates with CT033/SC02/GL01-08, which is composed of:

– UNI EN ISO 19650-1-2, *Information Management*;
– UNI EN ISO 16739, *Industry Foundation Classes* (IFC);
– UNI EN ISO 29481-1-2, *Building information modelling. Information delivery manual* (IDM), Part 1: Methodology and format, Part 2: Interaction framework;
– UNI-EN-ISO 12006-2-3, *Building construction—Organization of information about construction works* (IFD), Part 2: Framework for classification of information, Part 3: Framework for object-oriented information;
– ISO 16354, *Guidelines for knowledge libraries and object libraries*;
– ISO 16757-1-2, *Data structures for electronic product catalogues for building services*;
– ISO/TS 12911, *Framework for building information modelling (BIM) guidance*;
– ISO 22263, *Organization of Information about construction works: Framework for management of project Information*;
– ISO 10303, STEP.

With regard to Italy, the old **UNI 11337** regulation of 2009 "*Encoding criteria of construction works and products, activities and resources*" has been replaced with the new regulation "**Digital management of information processes for the building sector**" [13], of which parts 1, 3 (original), 4, 5, 6 and 7 have been published, while parts 2, 3 (new), 8, 9 and 10 will soon be published.

1.2   BIM's Foundations

In order to understand BIM in its current evolution, it is necessary to be informed on the two Anglo-Saxon international systems of reference, the British one: **UK**, and the American one: **USA**.
The British system is composed as follows:

- bodies,

  - *Royal Institute of British Architects (RIBA)*;
  - *UK BIM Task Force Group*;
  - *National British Standard (NBS)*.

- documents,

  - **BS-PAS 1192-2:2013** [12];
  - BS-PAS 1192-3:2014;
  - NBL National BIM Library;
  - BIM Toolkit;
  - CIC BIM Protocol:2013;
  - COBie—Construction Operations Building Information Exchange.

While the American system is composed as follows:

- bodies,

  - *American Institute of Architects (AIA)*;
  - *National Institute of Building Sciences—buildingSMART—(NIBS)*;
  - *US chapter of buildingSMART International (BIMforum)*;
  - *US Army Corps of Engineers (USACE)*.

- documents,

  - AIA G202-2013, Building Information Modeling Protocol Form;
  - AIA E203-2013, Building Information Modeling and Digital Data Exhibit;
  - AIA G201-2013, Project Digital Data Protocol Form;
  - National BIM Standard United States—V3:2013 (NBIMS-US);
  - National CAD Standard United States—V6:2014 (NCS-US);
  - **BIMforum LOD specification (2013–2019)**;
  - National BIM Guide for Owners;
  - BIM Project execution Planning Guide V2.1:2011;
  - USACE BIM contract requirements (UBR).

All these references put together allow to summarize some fixed elements concerning BIM that identify a common structure.

1.3 CAPEX and OPEX

BIM covers the entire useful lifecycle of products in the building sector, from strategy to demolition. PAS 1192-2 regulations identify the Development phase (CAPEX) and the Execution phase (OPEX).

CAPEX: strategy, feasibility, design, construction.
OPEX: management, maintenance, restoration, requalification, demolition.

Consequently, in PAS 1192-2 and 3, the models have both a regulatory function with regard to the production of products under construction—**Project Information Model** (PIM)—and an identification and registration function of the actual situation and of the passing of time with regard to existing products: **Asset Information Model** (AIM).

1.4 EIR and BEP

Information management requires the definition of an organized information flow, on the basis of three consequential phases.

1. The client requires information (Client, Employer, Appointing): Employer Information Requirement (EIR), a "document," or information container, that expresses the request for information and sets the rules for the dialogue between the contracting parties.
2. The tenderer puts in a tender relating to the management of information: BIM Execution Plan—pre contract—(pre BEP), "plan" for managing the information tendered on the basis of the client's requirements (EIR) and of the tenderer's standard abilities of execution.
3. The awarded party executes the information management: BIM Execution Plan—post contract—(post BEP), "plan" for managing the information agreed upon on the basis of the client's requirements (EIR) and of the tenderer's specific execution abilities verified and redefined on the basis of the specific contract and of the specific client.

For a long time, it was thought that the American structure consisted solely in the final awarded party's BIM job order (post-BEP). The USA structure, though, does not have a BEP (UK definition) but a BPxP (BIM Project eXecution Planning). Actually, the latter is not the executor's sole document, but part of a path that is identical (apart from the name) to the UK flow (Fig. 1.4). Therefore, going beyond the sole reading of Pennsylvania University's BPxP (PDF that can be freely downloaded from the Internet) [2], the USA structure is divided into three stages and does not go beyond the definition of the job order requirements, as many still think mistakenly [9].

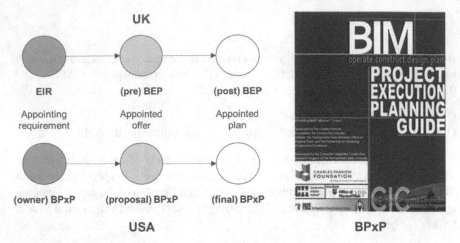

Fig. 1.4 UK and USA contract document flow

1. Owner—BPxP;
2. Proposal—BPxP;
3. Final—BPxP.

1.5   LOD

If CAD already highlighted that the concept of scale (of the design) had long lost much of its sense, with the exception of the printing phase of the Tables (2D), the introduction of BIM and the transit from design to simulation (virtualization, 3D) highlighted the need to introduce a new information measurement system. Therefore, the concept of LOD was developed as a measure of the quantity and quality of information provided. Today the acronym is taken for granted as well as the connection of the measure of the model's information (and of the information related to the building or infrastructure) with "objects" (physical elements). However, this has not always been the case and the actual acronym conceals different interpretations.

LOD was developed in the USA panorama (BIM Forum LOD Specification) as **Level of Detail** with reference to objects (door, wall, pillar, etc.). The term "Detail" continued to strongly connect the new system to the geometry of objects (geometrical representation). For this reason, in 2013 it was decided in the USA to introduce the concept of Development and the letter "D" became the initial letter of "Development:" **Level of Development**, with the aim to overcome the direct connection with geometry so as to consolidate the concept of information quantity and solidity. A high or low LOD has more or less information, but especially such information is more or less binding depending on the information evolution. The USA's LOD scale is measured in hundreds, from LOD 100 to LOD 500, where LOD 100 means that there are fewer and less consolidated data (such data can change when moving up to the following LOD and acquiring further in-depth information), compared to LOD 200 and so on. However, also the USA LOD requires the need to measure both the graphic geometrical information and the alphanumeric information. The BIM Forum Specification [1] is divided into: Part I—*Element Geometry*, and Part II—*Attribute Information*, even if the division has never been formalized.

Also, the British system developed the concept of information scale from the very outset: LOD, defined starting from PAS 1192-2:2013.

In this case, though, LOD means **Level of Definition**, which is divided into **Level of Detail**—LOD (for the graphic-geometrical attributes) and **Level of Information**—LOI (for the non-geometrical or alphanumerical attributes). Therefore, in the UK: LOD = LOD + LOI. Besides, between the UK and the USA the letter "D" has three meanings = development, detail, definition.

Moreover, in the UK system the LOD scale refers to the "*model*" in PAS 1192-2—changing name depending on the project stages (Brief, Concept, etc.), **Level of Model Definition** (Fig. 1.5):

– Since 2015, the UK's LOD also refers to the "*objects*" in the NBS BIM Toolkit, with the more renown measure in units (LOD from 1 to 6; Fig. 1.6).

To date, at least a dozen of LOD systems are counted worldwide, many of which use a scale in hundreds (LOD 100, 200, etc.) similar, but not identical, to that of the USA. In Italy, UNI 11337, part 4, introduced a structured system of LOD, referred to objects and measured in Letters (from A to G) so as not to confuse the market (since it is a synthesis of the UK and USA experiences brought within the Italian specificities) (Fig. 1.7).

The introduction of LOIN (Level of Information Need) and the overcoming of LOD through the new ISO 19650-1:2018 [6] may help find a synthesis between the various national standards (Fig. 1.8).

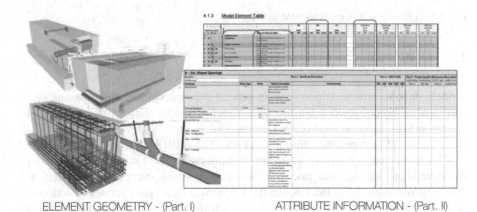

ELEMENT GEOMETRY - (Part. I)   ATTRIBUTE INFORMATION - (Part. II)

Fig. 1.5 LOD of BIM FORUM specification USA

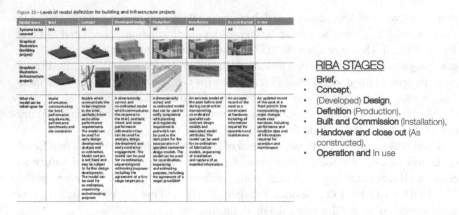

Fig. 1.6 Level of model definition of PAS 1192-2:2013

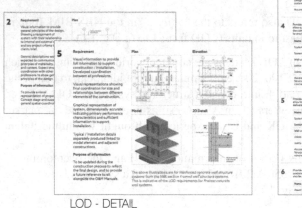

LOD - DETAIL   LOI - INFORMATION

Fig. 1.7 LOD of object in the NBS BIM toolkit

1.6   CDE

Models, data, records and documents in general in an information management system need to be collected according to a structured organization. The concept of Common Data Environment (CDE), introduced in the UK design regulation BSI 1192-1:2007 [11], is recalled digitally in PAS 1192-2:2013.

CDE is defined as the place where all the subjects involved in a specific job order can store, share, manage and process information in order to carry out their activities. The structure assumed in the UK standard system is based on 4 scopes of action—work in progress, share, publish, archive—in which the information transits through subsequent gates of approval (Fig. 1.9).

LOD A	LOD B	LOD C	LOD D	LOD E	LOD F	LOD G
Geometria Elemento architettonico rappresentato mediante un simbolo 2D.	**Geometria** Rappresentazione geometrica 3D del sistema facciata attraverso solidi con forma e spessore approssimati e linee d'asse che determinano la suddivisione della facciata in moduli.	**Geometria** Elemento architettonico verticale o pseudoverticale rappresentato con forma, dimensioni e posizione corretti.	**Geometria** Elemento architettonico verticale o pseudoverticale rappresentato con forma, dimensioni e posizione corretti, integrati da interfacce con altri sistemi.	**Geometria** Elemento architettonico verticale o pseudoverticale rappresentato con forma, dimensioni e posizione corretti. Sono rappresentati tutti gli elementi fisici che compongono la facciata e i componenti accessori. Sono definiti materiali, finiture e i dati specifici del fornitore di prodotti commerciali.	**Geometria** Come LOD D (con aggiornamenti rilevati in cantiere, se necessari).	**Geometria** Come LOD F.
Oggetto Grafica 2D (linee e campiture 2D)	**Oggetto** Solido 3D + linee d'asse	**Oggetto** Solido 3D composito + linee d'asse	**Oggetto** Solido 3D composito	**Oggetto** Solido 3D composito	**Oggetto** Solido 3D composito	**Oggetto** Solido 3D composito
Caratteristiche Posizionamento di massima	**Caratteristiche** Semplici geometrie d'ingombro	**Caratteristiche** Proprietà del pannello di facciata: • Tipologia di facciata • Definizione materiali • Dimensioni • Presenza di elementi	**Caratteristiche** Proprietà del pannello di facciata: • Tutte le caratteristiche del LOD C • Sistemi di fissaggio e altri elementi di inter-	**Caratteristiche** Proprietà del pannello o dei singoli componenti: • Tutte le caratteristiche del LOD D • Finiture • Tipologia del vetro	**Caratteristiche** Proprietà del pannello di facciata: • Tutte le caratteristiche del LOD E • Piano di manutenzione	**Caratteristiche** Proprietà del pannello di facciata: • Tutte le caratteristiche del LOD F • Data di manutenzione/ sostituzione

Fig. 1.8 Italian LODs of UNI 11337, example of curtain wall

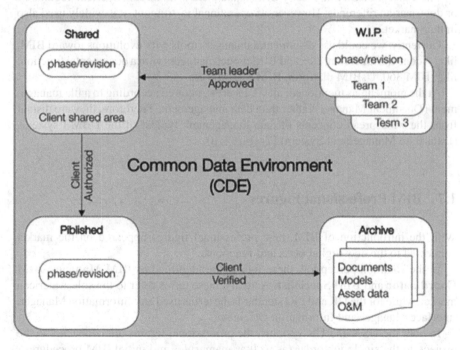

Fig. 1.9 CDE structure of BSI 1192 standard (anticlockwise flow)

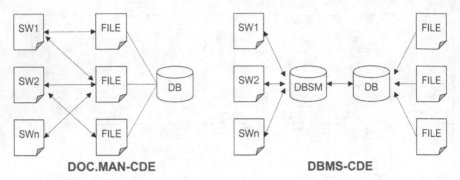

Fig. 1.10 CDE-Document manager and CDE-DB management system

Despite the name "Data Environment," also the most evolved systems currently in commerce concentrate their activity on the management of "files," although they are sensibly different from simple document or ftp managers.

The concept of CDE is strictly British and there is no direct American regulatory or doctrinarian reference. However, as operational instrument, it is widely used also in that market.

Currently, we could use documental managers tools with evolutions toward BIM, like: Aconex, Blue Beam, etc., and BIM model managers with a documental function, like: BIM 360, UsBIM platform, BIM+, BIMx, etc.

At the moment, as mentioned, all CDEs are structured according to a file management (Document Manager) rather than data management. Therefore, they are distant from the widespread concepts of data management typical of the DBMS systems (Data Base Management System) [7] (Fig. 1.10).

1.7   BIM Professional Figures

With the introduction of BIM, new professional figures appeared on the market connected to the new digital tasks and functions.

In the common perception, these figures are identified as **BIM Manager**, **BIM Coordinator** and **BIM Specialist**. Actually, these terms refer to the sole American market, while for the UK and ISO standards the terms used are: Information Manager, Interface Manager and Information Originator.

In both the USA and UK systems, the professional figures are connected to the project, to the single job order (as well as many rules and initial BIM procedures). Over the years, it has been increasingly necessary to think of higher professional figures, at organizational level (for the USA, Information Manager). However, none of the mentioned figures has a univocal and structured definition and framework at international level.

Table 1.2 BIM roles and professional figures in the world

Level	USA	UK/ISO	UNI
Organization	Information Manager	//	Gestore dei processi digitalizzati
	//	//	(Gestore Piattaforma Info)
Project	BIM Manager	(project) Information Manager	(Gestore p. d. di progetto)
	//	Task Information Manager	//
	BIM Coordinator	Interface Manager	Coordinatore flussi informativi di commessa
	BIM Modeler/Specialist	Information Originator	Operatore avanzato della gestione e della modellazione informativa
	//	(// CDE)	Gestore ACDat

The first regulation that defines BIM professional figures in a complete and structured way is UNI11337-7:2018 [14]. Using the American terms (translated also into Italian), UNI 11337 defines three managerial figures—BIM Manager, BIM Coordinator and CDE Manager (the first figure at organizational level, and the other two at project level)—and an operational figure, BIM Specialist (this too relating to the project). Therefore, compared to the USA original structure, the BIM Manager and the Information Manager are combined in a sole role and established at organizational level. The CDE Manager is introduced (CDE translated into Italian as ACDat) as a specific figure integrating the BIM AEC environment and the pure IT environment (Fig. 1.9 and Table 1.2).

1.8 BIM Dimensions

With regard to BIM, the first dimension of reference is the 3D simulation model. The third dimension rises from the two-dimensional Cartesian Coordinate System typical of the 2D traditional design.

Subsequently, its integration with the entire lifecycle of the products of the sector (buildings and infrastructure) highlighted further possible BIM "dimensions":

- timing, 4D (programming and scheduling of resources);
- cost, 5D (quantity take off and bill of quantity);
- operation, 6D (property e facility management);
- sustainability 7D (economic, social, environmental, social, etc.).

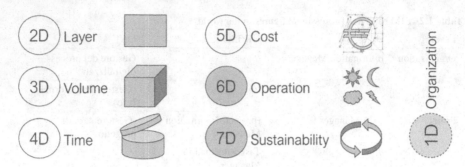

Fig. 1.11 BIM dimensions according to UNI 11337-1:2017

All the above without forgetting that BIM, as a method and not only as a tool, requires a coherent implementation mainly at 1D organizational level (definition of the 7 BIM dimensions according to UNI11337-1:2017) (Fig. 1.11).

1.9 BIM Levels

PAS 1192-2:2013 introduced the concept of levels of maturity of BIM or of its actual implementation in the building process.

Its evolutionary representation from CAD 2D (level 0) to iBIM (Integrated Building Information Model; level 3) ascribed to M. Bew has become renown (it cannot be reproduced because covered by copyright; see Fig. 1.12). As known, although interpreted differently, Level 2 is the one recalled in the minimum objectives of the UK governmental strategy up to 2025 (see par. xx).

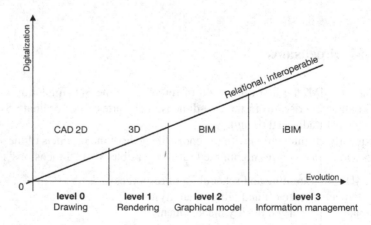

Fig. 1.12 BIM maturity level UK PAS 1192-2:2013

In PAS 1192-2, Level 2 of the BIM maturity is summarily identified in seven stages (from "a" to "g") (while nothing is said about the previous levels and their evolution to Level 3):

(a) *"originators produce definition information in models which they control, sourcing information from other models where required by way of reference, federation or direct information exchange;*
(b) *provision of a clear definition of the employer's information requirements (EIR) and key decision points (to form part of the contract possibly through adoption of the CIC BIM Protocol);*
(c) *evaluation of the proposed approach, capability and capacity of each supplier, and their supply chain, to deliver the required information, prior to contract award;*
(d) *a BIM execution plan (BEP) shall be developed by the supplier containing: (1) assigned roles, responsibilities and authorities; (2) standards, methods and procedures; and (3) a resourced master information delivery index, aligned with the project program;*
(e) *provision of a single environment to store shared asset data and information, accessible to all individuals who are required to produce, use and maintain it;*
(f) *application of the processes and procedures outlined in the documents and standards indicated in* Table 1.1 [M. Bew Schema];
(g) *information models to be developed using one of the following combinations of enabling tools: (1) discipline-based software, with individual proprietary databases, that have limited interoperability between them or with associated design analysis software; (2) discipline-based software, with individual proprietary databases, that are fully interoperable, but with limited interoperability with associated design analysis software; (3) discipline-based software, with individual proprietary databases, and associated design analysis software that are fully interoperable; single source platform software, with a single external relational database, and associated design analysis software that are fully interoperable."*

1.10 Conclusions

In order to understand BIM and to be able to use it profitably in its processes—explained in detail in the following chapters—it is important to analyze whether or not the market has currently absorbed the relevant novelties and up to what level, both as tool and as method. Even if "recent," the BIM phenomenon has already several interpretation stratifications that the market has difficulty in reading critically, believing instead that they have already been metabolized by users without the need for further in-depth analyses. Too many concepts are taken for granted and are disseminated and shared without verifying the true confidence of the subjects involved, or their common interpretation. If this is inappropriate for any practice or doctrine, it is even more so with regard to information management. The dissemination of a more structured BIM culture compared to the current free access on the Internet ("wikyBIM") and the dissemination of standards (ISO EN UNI) is fundamental in order to create a common basis of digital knowledge of reference that will allow the market to further evolve and advance. In other words, it is important to overcome the current trend at the basis of the dissemination of the thousands of single Proprietary Guidelines that risk to fossilize the method in thousands of bonds for the advantage of only a few consultants and large enterprises: *"…Without a standard data and process definition the supply chain and client will be re-creating a diverse range of proprietary approaches* [proprietary BIM Guidelines] *which will potentially add a cost burden for each project…"* (EUBIM. Task Group, par. 3.1.3, p. 48).

References

1. BIMForum (2015) Level of development specification. BIM Forum, p 195
2. Computer Integrated Construction Research Program (2010) Building information modeling execution planning guide—version 2.0. The Pennsylvania State University, University Park
3. Direttiva 2014/24/UE del parlamento europeo e del consiglio del 26 febbraio 2014 sugli appalti pubblici e che abroga la direttiva 2004/18/CE (no date)
4. Eastman MC et al (1974) An outline of the building description system. Carnegie-Mellon University, pp 1–23. Available at: http://files.eric.ed.gov/fulltext/ED113833.pdf
5. EU BIM Task Group (2016) Handbook for the introduction of building information modelling by the European public sector. Available at: www.eubim.eu
6. International Organization for Standardization (no date) ISO 19650-1, organization of information about construction works—information management using building information modelling—part 1: concepts and principles
7. Liang J, Chang TYP, Chan CM (1998) An object-oriented database management system for computer—aided design of tall buildings. Eng Comput 14:275–276
8. Lorimer J, Bew M (2011) A report for the government construction client group—building information modelling (BIM) working party strategy paper
9. Messner J, Anumba C (2010) Building information modeling project execution planning guide. Pennsylvania
10. Sutherland IE (2003) Sketchpad: a man-machine graphical communication system. Computer Laboratory—Technical Report, University of Cambridge, Cambridge, Available at: http://www.cl.cam.ac.uk/TechReports/
11. The British Standards Institution (2007) BS 1192:2007, collaborative production of architectural, engineering and construction information—code of practice. UK
12. The British Standards Institution (2013) PAS 1192-2:2013, specification for information management for the capital/delivery phase of construction projects using building information modelling. British Standard Institute, UK. ISSN 9780580781360/BIM task group
13. UNI/CT033/WG05 (2017) UNI11337-1:2017, Building and civil engineering works—digital management of the informative process—part 1: models, documents and informative object for products and processes. UNI, Milano, p 26
14. UNI/CT033/WG05 (2018) UNI11337-7:2018, Edilizia ed opere di ingegneria civile—Gestione digitale dei processi informativi delle costruzioni—Parte 7: Requisiti di conoscenza, abilità e competenza delle figure coinvolte nella gestione e nella modellazione inforamtiva'. UNI, Milano, p 19
15. van Nederveen GA, Tolman FP (1992) Modelling multiple views on buildings. Autom Constr. https://doi.org/10.1016/0926-5805(92)90014-b

Chapter 2
Standardized Structures for Data Collection According to Specific BIM Uses (Technical Datasheet)

Abstract The exploitation of potentialities offered by ICT (Information and Communication Technologies) can ease an efficient information management but to do that it is first of all necessary to use a standardized and univocal system for classifying, defining, collecting and archiving information related to the whole life-cycle of a building, from the initiative stage to its end of life. The aim of this chapter is to provide a proposal on how to structure such system, on the basis of an analysis of the state-of-the-art of the classification systems currently available. Such analysis led to define a new univocal classification and denomination system that structures each object (from the simplest—such as the construction product—to the more complex—such as the entire constructed facility or infrastructural work) within a very precise hierarchical scale. Moreover, standardized datasheets for data collection of construction products, in situ elements and assembled systems are described.

2.1 Introduction

The building sector is characterized by the presence of a multiplicity of languages that, in many cases, vary considerably according to the provenance and background. Therefore, the many criteria, methods, procedures, different languages and rules adopted in the building sector increase the existing difficulties to exchange information among the different subjects [1]. In fact, the different stages of the building process involve a plurality of actors that generates too often issues mainly related to a not very efficient information management. Such misunderstandings can cause a low building quality, a considerable waste of time and economic resources. This in order to allow an optimized use of data, so as to enable the single actors of the building sector to effectively and efficiently use the developed tools.

The aim of this chapter is to provide a proposal on how to structure such system, on the basis of an analysis of the state-of-the-art of the classification systems currently available (OmniClass [2], Uniclass [3], PC/SfB [4], RUDC, UNI 8290-1:1981 [5], IFC [6], UNI 11337:2009 [7]). Such analysis led to define a new univocal classification and denomination system that structures each object (from the simplest—such

© Springer Nature Switzerland AG 2020   17
B. Daniotti et al., *BIM-Based Collaborative Building Process Management*, Springer Tracts in Civil Engineering,
https://doi.org/10.1007/978-3-030-32889-4_2

as the construction product—to the more complex—such as the entire constructed facility or infrastructural work) within a very precise hierarchical scale. The adopted logic is the following: each element is considered as an object, regardless of its complexity, and all the so-called information attributes are associated to each element. Activities and human resources were treated in the same way in order to standardize all information flows, included the managerial and procedural ones, allowing to archive and manage the related data.

Thanks to a database which is structured according to standardized technical datasheets and continuously implementable through additional information attributes, it is possible to collect all the information in a single tool. Stored data can then be used by the different actors of the construction sector through a duly studied web portal [8], through some web services which allow interoperability with plug-in, linked software and other databases.

In order to achieve such results, it is fundamental to define an ordered and univocal structure of information, necessary to describe each object, actor and action of the building process. In fact, at any level of development, a product of the building sector and its components (thus considering not only a facility as a resulting product, but also its spaces, systems, elements and composing construction products) are:

– objectified by a physical and tangible element;
– represented by one or more information and intangible entities.

Therefore, the management of the building process and its final product does not involve only material and physical aspects (e.g. to build, to carry out maintenance, to supply), but also intangible and informative aspects nature (of knowledge) [9].

In order to guarantee an adequate information management during the building process, such information has to be defined in its characteristics, understandable in its meaning and transferable among each actor who is participating at any title. Therefore, an efficient exchange of information needs the definition of criteria capable of identifying and describing in an univocal manner the actors and objects that intervene in the entire lifecycle of the work. Once these criteria have been defined, it is fundamental to implement ICT tools to be used along the building process, in such a way to make the information easily accessible and usable by the different subjects involved.

For this reason, the INNOVance platform represents the first national database for the building sector capable of archiving, updating, transmitting in a clear, standardized and interoperable way all the information of the construction sector. It is based on:

• a standardized system for the classification and denomination of the objects, actors and activities of the construction sector;
• a standardized system for the collection and archiving of the data related to the mentioned objects, actors and activities;
• a regulated system for the access and transfer (ingoing and outgoing) of the above mentioned data.

On one hand, the INNOVance database was developed in such a way to structure all the information on construction products efficiently: it considers information related to the facility, plant-engineering, means or equipment, so as to facilitate choices and provide high transparency with regard to the completeness of the information provided by manufacturers and by the involved actors. On the other hand, such database also allows to store the information relating to the building process, allowing the involved professionals to read and download the most updated information falling within their interest and to download and share documents or data falling within their competence.

The intent is to considerably facilitate the identification and choice of products, materials and solutions to be adopted, and to provide all relevant information in a transparent and immediately usable way.

2.2 Information Management

The information management system developed in INNOVance considers the constructed facility according to four main logics:

- functional-spatial logic: it allows to highlight the functional spaces and areas associable to a specific work [10];
- technological logic: it allows to disaggregate the constructed work in its technological components;
- environmental-anthropic logic: it allows to represent the series of changes to the environment and the landscape-natural aspects associable to the constructed work;
- procedural logic: it allows to describe in a complete way all the activities, tools, vehicles and human resources necessary to realize a facility or part of it.

Construction and information flows are therefore correlated and disarticulated in such a way to break the work down into functions and objects:

- when taking into account functions, the functional-spatial system is considered;
- when taking into account objects, the considered systems are the technological, the environmental-anthropic and the procedural ones.

Figure 2.1 shows the distinction of the work between those functions and objects composing it and the systems taken into consideration on the basis of families of objects that are useful to explain the methodology according to which the constructed work is divided.

The following paragraphs focus the attention on the technological system of buildings, highlighting the reasons for the development of the classification and denomination system and the criteria for the definition of the technical datasheets of the construction products, the in situ elements and the assembled systems.

In particular, the technological system of buildings is the structured set of technological units and/or technical elements defined in their technological requirements and in their specifications of technological performance.

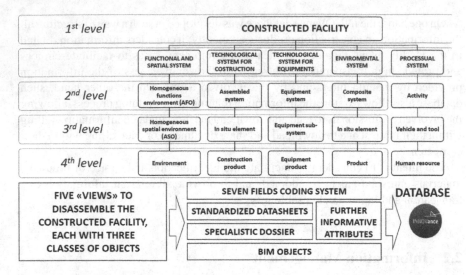

Fig. 2.1 INNOVance hierarchical structure [11]

The technological system of buildings is structured on the basis of 3 levels:

- construction product;
- in situ element;
- assembled system.

The logic followed in order to structure the levels is technological: the work, in fact, is viewed as a formalization of various technological systems, composed—increasing the object complexity—of construction products, in situ elements and assembled systems. Construction product means any manufactured product placed on the market to be incorporated in facilities or in parts of them and whose performance affects the performance of the constructed facilities with regard to their basic requirements.

The in situ element is a construction product supplied and installed. It is different from the industrial product due to the presence of one or more manufacturing processes, it accomplishes a characterizing function and is part of an assembled system, carrying out (or contributing to carry out) one or more functions.

The assembled system is the more or less structured composition of in situ elements combined among each other due to the common correspondence to an aggregating function. This is the result of one or more works correlated among each other with the aim to obtain a complex, functional and/or physical spatial (tangible) system.

For each object belonging to the three levels, besides a standardized classification and denomination to identify families of objects with common characteristics, it is possible to provide precise information through standardized parameters collected in technical datasheets and further information attributes.

The information describing precisely the construction products was obtained also owing to a cooperation with various category Associations and Federations of manufacturers of products and materials relating to the building sector, participating in a specially devoted technical table within UNI. In line with what provided for by the national technical regulations of reference and with what is in phase of development, and considering the current market needs, the system developed allows to univocally name the products and standardize the collection of information relating to their lifecycle. Through the technical datasheets drawn up by various producers and uploaded in a specially devoted section on the INNOVance portal, it is possible to know and compare the technical and performance characteristics of products and make use of project suggestions for installation, maintenance and management.

The structure defined for collecting information on construction products was then widened and adapted to store data also regarding in situ elements and assembled systems.

2.3 The Univocal Classification and Denomination System

Due to the multiplicity of involved actors, languages, criteria, methods and habits typical of the building sector, it is very important that the transmission of information can flow simply and, especially, avoiding misunderstandings. The process of communicating information among different actors within the construction sector constitutes a critical factor that must be correctly faced: it is, first of all, fundamental to be able to identify unequivocally a construction product, an activity, a performance, a measurement and to communicate these kind of data according to shared rules, in an efficient (reduction of timing and costs) and effective way (correspondence of the final output to the technical specifications required for the project).

To provide such approach with a single communication tool it is necessary to adopt standardized ways of describing constructed works and their parts; a shared classification and denomination system is a means to achieve such objective, facilitating data exchange among different subjects in the same sector.

A classification system must have two essential characteristics in order to be considered effective: it must define a topic exhaustively including all related concepts and it must allow for each element to belong to a single class. Therefore, a classification system must meet the following two requirements:

- it must be stable because once it has been disseminated it must be applicable to the different contexts envisaged without needing any substantial modifications or radical changes;
- it must be flexible because it must guarantee the possibility to be expanded so as to allow, when necessary, to add new parts; as a consequence, it must be conceived with a modular structure.

The study of the classification approaches adopted in the building sector highlights a clear difficulty to classify products univocally. It also emphasizes a great fragmentary nature of these classifying systems, such to make them usable only locally. Exceptions to the local classifying systems are SfB and the RUDC system, which was not conceived directly for the building sector because it is a reduced version of the UDC plan (Decimal and Universal Classification), conceived to organize all sort of information.

The basis of a classification system is the division of a series of elements into classes; there are two different modalities to group elements:

- direct grouping, when the elements are identified as belonging to a class and the classes are organized according to a hierarchical order. Therefore, there are core classes and, for each one of them, there are subclasses and so on (for example, the parts of a building are walls, floors, foundations, roof, etc., and within such groups it is possible to identify other elements); an example of a classification based on direct grouping is CSI Master Format [12] widespread in the USA.
- combined grouping, when several attributes are considered for one element, which is therefore identified on the basis of the free aggregation of attributes. A classification based on this type of grouping is called faceted. Examples of a faceted classification are SfB and the Italian regulation UNI 8290.

2.4 State-of-the-Art of Classification Systems

In Italy price lists are traditionally used to classify construction elements univocally in order to indicate average prices of different building works and activities (price lists of the Chambers of Commerce, price lists of the Civil Engineers, etc.). In these price lists, the conventional works referring to the worksite are classified according to a hierarchical scheme, so one or more products with common characteristics are assigned to the single manufacturing process indicating:

- product macro-categories;
- product classes;
- manufacturing (or conventional items).

This allows to guarantee a direct correspondence between the elements of the classification and the manufacturing process necessary to realize the physical contents provided for by the construction project.

UNI 8290 represents an important step forward in providing an official and widespread national classification model, with wide acknowledgment and testing in the professional and institutional practice. In particular, part 1 was drawn up with the aim to allow an ordered and structured division of a building system into various levels, with homogeneous rules. The system is divided into three levels, as follows:

- classes of technological units (first level);
- technological units (second level);
- classes of technical elements (third level).

The items of the first two levels are the most suitable to represent functions aimed at satisfying users' needs. The items of the third level correspond to classes that provide an overall or partial answer to the functions of the technological units.

Although such classification is referred to residential buildings, it can also be used for facilities with different types of use. For operational purposes, it is possible to extend it to further levels, the important is that they can identify increasingly each constituting component and that they are homogeneous among each other.

The SfB classification plan (acronym of the Swedish Samarbetskommittén for Byggnadsfrägor) is less used in Italy. It was developed in 1956 with the aim to become a system adoptable at international level. Realized in Sweden, it was presented in the Netherlands in the presence of many experts coming from all over Europe. The system had great success and the first English version was translated into German and Italian. Basically, the method makes use of various sets, or groups, or encoding levels called "tables," within which a code is identified for all fundamental elements connected to the building process. Table 1—the Elements Table—and Table 2—the Shapes/Materials Table—are the most important. This method is organized on the basis of three sections: (1) the section of the composing elements (the designer's view, very close to the logic of the functional classification); (2) the section of the construction activities (the builder's view); (3) the section of the materials and resources, incorporated in the buildings. These different sections correspond to just as many schemes which are: no. 1 (Elements); no. 2 (Activities); no. 3 (Materials).

Starting from UNI 8290 and from the SfB classification, the working group "Autorità-Ance-Dei" developed a study that resulted in UNI 11337 "Building sector and civil engineering works – Encoding criteria of construction works and products, activities and resources – Identification, description and interoperability" ["*Edilizia e opere di ingegneria civile - Criteri di codificazione di opere e prodotti da costruzione, attività e risorse - Identificazione, descrizione e interoperabilità*"]. The purpose of the work was to draw up a list of items according to the logic of constructed facilities, in order to allow analytical estimates (such as the cost per unit according to the calculation of the single interventions necessary to realize a work) and elementary estimates (such as the cost per unit according to the calculation of the technological unites).

Whereas, among the classification systems which are very well known abroad there is Uniclass (Unified Classification For The Construction Industry). Introduced

Table 2.1 Comparison between the mentioned classification systems

System	Function	Object/material	Process/manufacturing
PC SfB	x	x	
UNI 8290	x		
UNI 11337	x	x	x
UNICLASS		x	
OMNICLASS	x	x	
IFC		x	

in the United Kingdom in 1997 by the Construction Project Information Committee (CPIC), it proposes a method for classifying the construction sector on the basis of 15 tables; each table represents a specific information aspect and can be used singularly or combined with other tables to express complex concepts.

MasterFormat is the most used communication standard in the USA and Canada that allows to organize the contents of the project and related documents. The project, through a proposal submitted by the Construction Specification Institute (CSI), was divided into divisions and within each division into sections, with the aim to standardize a procedure capable of managing all the information connected to the project.

OmniClass is a classification system adopted by the American construction industry (2006); it is an open classification that can be used for free. It is based on existing systems still in use, such as Uniclass, MasterFormat and UniFormat [13]. At the basis of the system proposed by OmniClass there is the research of a common language to be adopted by the entire building sector providing a standardized system to classify information concerning the whole lifecycle of a generic construction, from conception to demolition. Another purpose is to guarantee coherence within the information system in all the planning phases, owing to the structured identification of the subjects of the project and the identification and management of the relations among them.

The logics adopted in the mentioned classification systems are summarized in Table 2.1.

2.5   INNOVance Classification and Denomination System

Starting from the analysis of the classification systems mentioned above, the classification and denomination system proposed by INNOVance allows to identify the work and all of its parts/portions throughout all phases of the production process. In fact, not only is it possible to identify the components of the work, but also the results of the worksite activities that in turn generate outputs.

Given the complexity of the context of reference, the process for establishing the classification system utilizes a rational approach, divided into 5 main stages:

1. monitoring of the environment: analysis of the building sector;
2. identification of the problem: analysis of the sector's problems with regard to the applicable encoding systems with consequent understanding of their limits;
3. identification of objectives: development of a univocal encoding system for the sector;
4. identification and evaluation of alternative solutions: three possible encoding systems were identified, evaluated on the basis of fixed objectives;
5. choice of the best alternative: the system selected was considered the best, however highlighting the presence of possible and often unavoidable trade-offs.

The new rules of the system were defined, first of all, by making choices, even difficult because in contrast with a multiform procedure to which the professionals pretended to uniform over time. In order to carry this out, the basic characteristics of the process were redefined, the pieces necessary for its functioning were identified and the relevant meanings were encoded in order to reconstruct on such basis a common language capable of working as a common denominator for the professionals of the sector.

Once defined the general rules, on the basis of the classification systems analyzed, a path was chosen which, in the rigor of a standardized structure, could guarantee a flexible use such to embrace all possible fields of use, in a certainly wider panorama than the one that every single production sector has before it when deciding to standardize its single information flow.

In fact, the classification system defines a structuring bound to an open editing, composed of seven fields, besides a further field that functions as a code to identify the macro-category of belonging (e.g. building product, in situ element or assembled system). This structure is analogous both for the construction and for all the objects defined in the levels in which the construction is disassembled.

2.6 Application to Construction Products

With regard to construction products, the seven characteristics for their denomination are as follows:

1. category: the field identifies homogeneous construction families on the basis of function and performance (masonry element, window, door, wooden element for internal flooring, concrete with guaranteed performance, prefabricated structural element);
2. typology: the field diversifies the class of the construction product, providing indications on the peculiar typological character (block for collaborating floor, continuous façade with uprights and beams, door resistant to swinging door forcing);
3. regulatory reference: the field indicates the technical specification harmonized for EC marking, when present (for example, product for discontinuous coverings UNI EN 1304, window UNI EN 14351-1, thermal isolation UNI EN 13162); otherwise, it identifies a possible regulation or guideline;
4. performance characteristics: the field indicates the prevailing performance of the construction product (element for masonry with thermal conductivity equivalent to = ... W/(mK), masonry mortar with guaranteed performance ...);
5. geometrical characteristics: the field provides information of various nature on shape, geometry, packaging, etc. (thermal insulating panels, window with a double rectangular shutter);
6. dimensional characteristics: the field indicates dimensions (floor block with height = ... mm, length = ... mm and width = ... mm, concrete with guaranteed performance with maximum dimension of the aggregate = ... mm);

7. physical-chemical characteristics: the field indicates the composition of the matter (door with wooden frame and glass stopping, profile for aluminum doors).

At terminological and semantic level, the terms used in INNOVance are derived from the national and community regulations in force. Moreover, an actual lexicon of reference for the building sector was drawn up, with the aim to provide a univocal language shared by all operators. In order to foster communication and the exchange of information within the sector, the lexicon also contains a collection of certified synonyms in common use [14, 15]. The seven-field structure was adopted also for the denomination of in situ elements and assembled systems, for which it was decided that it would have been more useful to provide the functional characteristics instead of the regulatory reference.

2.7   Application to In Situ Elements

The main characteristics that identify in situ elements are as follows:

1. category: the field identifies the family of in situ elements, whose specification is structured in the following fields;
2. functional characteristics: the field identifies the functions and scopes of use provided for the in situ element. This field, for example, indicates the belonging of the in situ element to a closing, a division, etc.;
3. typology: the field allows to diversify the class of in situ elements, providing indications on the peculiar typology;
4. performance characteristics: the field identifies the prevailing performance of the in situ element;
5. geometrical characteristics: the field allows to provide information concerning the shape, the orientation in space, the aesthetic and construction aspects relating to the in situ element;
6. dimensional characteristics: the field identifies the dimensions necessary for the unequivocal denomination of the in situ element;
7. physical-chemical characteristics: the field indicates the matter that allows to identify the construction product that characterizes the most the in situ element to which the product belongs.

2.8   Application to Assembled Systems

The main characteristics that identify the assembled systems are as follows:

1. category: the field identifies the family of assembled systems, whose specification is structured in the following fields;

2. functional characteristics: the field identifies the functions and scopes of use provided for the assembled system. This field, for example, indicates the belonging
   of the assembled system to a closing, a division, etc.;
3. typology: the field allows to diversify the class of assembled systems, providing
   indications on the peculiar typology, such as for example whether or not there is
   thermal isolation, acoustic isolation, waterproofing, etc.;
4. performance characteristics: the field identifies the prevailing performance of the
   assembled system;
5. geometrical characteristics: the field allows to provide information concerning
   the shape, the orientation in space, the aesthetic and construction aspects relating
   to the assembled system;
6. dimensional characteristics: the field identifies the dimensions necessary for the
   unequivocal denomination of the assembled system;
7. physical-chemical characteristics: the field indicates the matter that allows to
   identify the in situ element that characterizes the most the entire assembled
   system.

2.9 Standardized Collection of Information

In order to fully describe the object (a construction product, an in situ element, an
assembled system or an entire constructed facility), besides the univocal denomination, many information attributes are necessary, collected in technical datasheets
and referring to different aspects (technological, project, maintenance, economic,
operational aspects or other). Therefore, the technical datasheet represents a virtual
scheme of a rational and organized collection of information relating to a specific
subject, object or activity of the building process [16].

Within the INNOVance project, the attributes are grouped according to two levels:

- technical information for the description of the actual object and for its correct
  supply/installation, use, maintenance and demolition;
- further attributes for evaluations on security, environmental certifications, analysis
  of resources.

Some attributes are static, because the provided data concern a specific "standard"
object or they describe it once its realization has been completed; some other attributes
instead have to be continuously updated during the lifecycle of the considered object.

The standardization of information allows to group data according to homogeneous criteria and sections. In such a way, univocal information allows the professionals of the sector to make a quick comparison between the products belonging to
the same category, manufactured by different companies, facilitating the decisions
in each stage of the building process. The comparison between products can be carried out according to different points of view. Depending on the section to which
the data to be investigated belong, it is possible to see which technical datasheets

are enhanced with a greater quantity of information compared to those that show a lacking part.

When there are no common and homogeneous guidelines, the descriptions of the construction products are provided autonomously for each product sector of reference, with reference to the specific regulations in force. Therefore, it results difficult to have an exhaustive description of an object and to compare objects of the same nature among each other. With the reorganization of the information flow, capable of keeping trace also of the continuous regulatory updates, as well as the behavior of the products within their lifecycle, a system for the complete collection of information is developed relating to different aspects, such as use, maintenance, sustainability, etc.

The data collection and archiving takes place starting from formats of technical datasheets that, through a common language, can be univocally understood by all subjects in the building sector. In order to be efficient, the datasheets must aim at an ordered structure that allows to swiftly compare the different products, facilitating the choice, thanks to the transmission of clear information capable of meeting the designer's and client's different needs.

The format was developed considering the two regulations of reference for the technical regulatory framework in the construction sector:

- standard UNI 8690-3:1984 [17];
- standard UNI 9038:1987 [18].

The models for the collection of the technical information are organized in information blocks of homogeneous data. Such structure groups the multiplicity of the information in homogeneous classes and provides indications for editing the technical datasheets, standardizing their description and characterization process.

2.10 Standardized Technical Datasheet of Construction Products

The technical datasheet of construction products was developed within the scope of the Italian standardization body UNI, working group GL 9 "Encoding of construction products and processes"—"*Codificazione dei prodotti e dei processi in edilizia*"). With regard to construction products under CE marking, such technical datasheet is divided according to the following information blocks:

1. Information identifying the manufacturer

    1.1. Company name
    1.2. Website
    1.3. Registered office
    1.4. Manufacturing plant
    1.5. Contact
    1.6. Company certifications

2. Information identifying the product

    2.1. Denomination (according to UNI 11337)
    2.2. Identifying code (according to UNI 11337)
    2.3. Commercial name
    2.4. CPV code
    2.5. Other internal codes given by the manufacturer
    2.6. Intended use (obtained from the harmonized technical specification)
    2.7. Harmonized technical specification (hEN-EAD): denomination, classification, definition, code, number and year of regulation
    2.8. Description for tender specifications
    2.9. Description for the price list
    2.10. Synonyms
    2.11. Key words

3. Technical information

    3.1 Morphological-descriptive characteristics
        3.1.1. Geometry and shape
        3.1.2. Visual and constructive aspect
        3.1.3. Dimensions
        3.1.4. Physical-chemical: qualitative, quantitative
        3.1.5. Main components of the product
    3.2. Performance characteristics
        3.2.1. Essential characteristics
        3.2.2. Voluntary characteristics
    3.3. Information on sustainability
    3.4. Information on safety

4. Information on packaging, handling, storage in factory and transport

    4.1. Packaging
    4.2. Type of handling
    4.3. Storage modality
    4.4. Transport modality

5. Commercial information

    5.1. Average delivery time
    5.2. Commercial unit of measurement
    5.3. Product yield
    5.4. Insurance cover

6. Additional technical information
7. Complementary documentation

    7.1. Guidance dossier
    7.2. Technical datasheets of product components

8. Annexes

    8.1. Statements, certificates and authorizations: performance statement, voluntary certificates of the product, homologations

    8.2. Safety datasheet

    8.3. Graphic and multimedia annexes: photos, videos, designs, graphic details, etc.

    8.4. Further documentation

9. Information on data reliability

    9.1. Date of realization of the technical datasheet

    9.2. Information identifying the compiler

    9.3. Date of review of the technical datasheet

    9.4. Information identifying the reviewer.

The information blocks provided by the model for the collection of the technical information are characterized by main items, organized internally according to a criterion of further detail, differentiated depending on the fact whether they are products with or without the CE marking.

In fact, a similar structure was adopted also for the technical datasheets of products not yet subject to CE marking, with some minor differences with regard to regulatory references and performance characteristics.

2.11 Guidance Dossier

A guidance dossier was drawn up containing information on the supply, installation, the correct use, maintenance, and the modalities of demolition, according to the following information blocks:

1. Information identifying the product

    1.1. Manufacturer name

    1.2. Denomination (according to UNI 11337)

    1.3. Identification code (according to UNI 11337)

    1.4. Actual use

2. Information on transport, handling and storage

    2.1. Transport modality

    2.2. Type of handling

    2.3. Storage modality

    2.4. Rules on packaging disposal

3. Commercial information

    3.1. Sales network

4. Laying and installation
   4.1. Other construction products used for supply/installation
   4.2. Supply/installation
       4.2.1. Technical specifications
       4.2.2. Fit correlated materials and products
       4.2.3. Incompatible correlated materials and products
       4.2.4. Type of handling from the storage area to the working area
       4.2.5. Preconditions necessary for the supply
       4.2.6. Rules of use
       4.2.7. Modality of supply
       4.2.8. Acceptance criteria
       4.2.9. Indications on waste disposal
       4.2.10. Further specific requirements
   4.3. Use and maintenance
       4.3.1. Indications on maintenance
       4.3.2. Reference service life
   4.4. Demolition
       4.4.1. Dismantling/demolition modality
       4.4.2. Indications on reuse
       4.4.3. Indications on recycle
       4.4.4. Indications on disposal of "contaminated" element
       4.4.5. Indications on disposal of element in dumping ground
   4.5. Prevention and safety
       4.5.1. Safety in supply
       4.5.2. Safety in use
       4.5.3. Safety in maintenance
       4.5.4. Safety in demolition

5. Additional technical information
6. Complementary documentation

   6.1. Technical datasheets of the product

7. Annexes

   7.1. Graphic and multimedia annexes: photos, videos, designs, graphic details, etc.
   7.2. Further documentation

8. Information on data reliability

   8.1. Date of the completion of the guidance dossier
   8.2. Information identifying the compiler
   8.3. Date of review of the guidance dossier
   8.4. Information identifying the reviewer.

Moreover, the database was completed with information not related only to production, such as for example:

- economic attributes, such as: price according to price list, purchase price, average discount, price of supply;
- attributes connected to safety, such as: H phrases (hazard indications), P phrases (prudence recommendations) and combinations of P phrases;
- attributes connected to the management of packaging waste, such as: CER code, disposal code, recovery code.

After defining the contents of the technical datasheet and of the standardized guidance dossier to be used for the construction products, a specific layout was developed. Moreover, thanks to the contribution of the manufacturers' category associations involved in the INNOVance project, it was possible to draw up such standardized format for families of construction products falling within their competence [19].

2.12   Standardized Technical Datasheet of In Situ Elements

Analogously to what developed for the construction products, the functional layers information was grouped in macro-categories, in order to allow a swifter identification of the datum within the actual datasheet.

1. Information identifying the compiler

   1.1. Denomination/business name
   1.2. Registered office
   1.3. Website
   1.4. Contact

2. Information identifying the in situ element

   2.1.   Denomination of products
   2.2.   Denomination (according to UNI 11337)
   2.3.   Identifying code (according to UNI 11337)
   2.4.   CPV code
   2.5.   Intended use
   2.6.   Number of construction products for the realization of the in situ element
   2.7.   Description from technical tender specifications
   2.8.   Description from price list
   2.9.   Test reports
   2.10. Synonyms
   2.11. Key words

3. Technical information

   3.1. Morphological-descriptive characteristics
        3.1.1 Geometry, shape, visual and constructive aspects
        3.1.2 Dimensions

       3.1.3 Physical-chemical: qualitative, quantitative characteristics
       3.1.4 Tolerances
       3.1.5 Composition
  3.2. Performance characteristics
  3.3. Possible aspects relating to safety
  3.4. Economic programming
  3.5. Operational programming

4. Graphic representation in section
5. Additional information
6. Annexes

  6.1. Statements, certificates and authorizations
  6.2. Graphic and multimedia annexes: photos, videos, designs, graphic details, etc.
  6.3. Further documentation.

Although the structure derives from the one defined at the UNI working group for construction products, it was necessary to adapt various sections, deleting some and adding others.

The part relating to the performance characteristics was structured, for example, grouping the possible performances in the 7 basic requirements of CPR's works:

1. mechanical resistance and stability;
2. safety in case of fire;
3. hygiene, health and environment;
4. safety and accessibility with regard to use;
5. protection against noise;
6. energy economy and heat retention;
7. sustainable use of natural resources.

2.13 Standardized Technical Datasheet of Assembled Systems

With regard to the entire technological package, which in the INNOVance project is called assembled system, a technical datasheet was structured to allow an easy visualization of the its stratigraphic composition. Analogously to what developed for the in situ element, it indicates the performance characteristics grouping them on the basis of the basic construction requirements.

The information blocks of the assembled system datasheet are as follows:

1. Information identifying the compiler

  1.1. Denomination/business name
  1.2. Registered office

    1.3. Website

    1.4. Contact

    1.5. Corporate certificates

2. Information identifying the assembled system

    2.1. Denomination of the assembled system

    2.2. Identification code, denomination, category and typology according to UNI 11337

    2.3. CPV code

    2.4. Planned use

    2.5. Number of in situ elements for the development of the assembled system

    2.6. Description from the technical tender specifications

    2.7. Description from the price list

    2.8. Test reports

    2.9. Synonyms

    2.10. Key words

3. Technical information

    3.1. Morphological-descriptive characteristics

        3.1.1. Composition

        3.1.2. Geometry, shape, visual and constructive aspect

        3.1.3. Dimensions

        3.1.4. Physical-chemical properties: qualitative, quantitative

        3.1.5. Tolerances

    3.2. Performance characteristics

4. Graphic representation in section

5. Safety aspects

    5.1. Operational programming

6. Economic aspects

    6.1. Economic programming

7. Managerial aspects

    7.1. Maintenance programming

    7.2. Insurance cover

8. Annexes

    8.1. Statements, certificates and authorizations

    8.2. Graphic and multimedia annexes: photos, videos, designs, graphic details, etc.

    8.3. Further documentation

Fig. 2.2 Web platform INNOVance: BIM library, view of the assembled system datasheet called "not load bearing brick wall" [20]

9. Information on data reliability

    9.1. Date of the completion of the technical datasheet

    9.2. Information identifying the compiler

    9.3. Review date of the technical datasheet

    9.4. Information identifying the reviewer.

The technical datasheets offer all operators the possibility to evaluate the performances of different technical solutions. Among specified data there are the thermophysical properties that define the thermal performance of the assembled systems, of the in situ elements or of the construction products, useful for energy efficiency analyses.

The INNOVance database can store all the compiled technical datasheets in the archive called BIM Library where they can be associated to BIM objects (Fig. 2.2).

2.14 Use of Data

All the information provided is collected in a database of SAP's management system (SAP NetWeaver). For a better drawing up and consultation of data, all the information is accessible and usable from a free web portal through which:

• manufacturers can create and modify the technical datasheets of the construction products;

- designers can fully describe the technical solutions designed, with the possibility to compare different construction products to be used in the technical solutions and to consult indications related to the correct laying and installation;
- users can consult in any moment the datasheets drawn up by manufacturers, designers, construction companies, checking the correspondence of the products with orders and arrivals at work sites.

Depending on the authentication data of the various users, the portal is provided with a private data section and a public data section. In fact, sensible data are not public and can be accessed only by the owner, who can decide to make them publicly visible or share them single other INNOVance users. The private part of the database constitutes an advantage for the owner that decides to utilize this database since it allows to have at disposal, swiftly and inexpensively, statistics on the owner's products, enabling to be more competitive on the market and to have greater visibility. In fact, INNOVance's portal is not addressed only to manufacturers, but also to suppliers, designers, construction companies, clients and managers of real estate assets.

In order to ease interoperability between software (with particular reference to software for energy modeling and BIM Authoring Tools), specific web services and plug-ins were developed in order to connect the information collected in the database accessible from the portal and the databases of several software (e.g. Autodesk Revit) [21]. This allows to guarantee the connection between the information in BIM models and the information that can be inserted and downloaded from the portal [22, 23].

2.15 Conclusions

Considering the difficulties in communicating and transmitting information identified in the building process, the INNOVance project has developed an integrated information management system, based on:

- the use of a univocal language;
- the standardization of the information and the drawing up of technical datasheets;
- the collection of all information attributes for the description of a work and of all objects composing it;
- the archiving of information of different nature in a central database;
- the interoperability between IT tools;
- the use of the information through a portal, with a simple and effective interface.

Starting from the use of a common language and a standardized structure for the collection of information, the entire building industry is likely to benefit from enormous advantages, such as the decrease in the waste of time and economic resources that is currently characterizing the building sector.

References

1. Daniotti B, Lupica Spagnolo S, Pavan A (2013) Un linguaggio univoco per l'edilizia. The unambiguous language for construction. In: ISTEA conference
2. OmniClass: a strategy for classifying the built environment
3. CPIC—Uniclass2
4. Clegg D, Ray-Jones A (1991) CI SfB construction indexing manual
5. UNI (1981) UNI 8290-1: Edilizia residenziale - Sistema tecnologico - Classificazione e terminologia
6. International Organization for Standardization (2013) ISO 16739:2013—industry foundation classes (IFC) for data sharing in the construction and facility management industries
7. UNI (2009) UNI 11337: Edilizia e opere di ingegneria civile - Criteri di codificazione di opere e prodotti da costruzione, attività e risorse - Identificazione, descrizione e interoperabilità
8. Pavan A, Daniotti B, Lupica Spagnolo S, Caffi V (2015) INNOVance anche per l'edilizia scolastica: la rivoluzione dell'information management. Costr Laterizio 160:66–71
9. Pavan A, Daniotti B, Re Cecconi F, Lupica Spagnolo S, Maltese S, Chiozzi M, Pasini D, Caffi V (2014) Gestione informativa delle costruzioni, INNOVance per il processo costruttivo. Construction Information Management (CIM), INNOVance for the construction process. In: ISTEA conference
10. Gottfried A, Pavan A, Chiozzi M, Devito AC, Pasini D, Carrante GF (2014) Flusso informativo per la gestione integrata della commessa. Information flow for the integrated procurement management. In ISTEA conference
11. Lupica Spagnolo S (2017) Information integration for asset and maintenance management. In: Sanchez AX, Hampson KD, London G (eds) Integrating information in built environments. From concept to practice, Taylor & Francis Group, Routledge, pp 133–149
12. MasterFormat [Online]. Available: http://www.csinet.org/masterformat
13. Charette RP, Marshall HE (1999) UNIFORMAT II elemental classification for building specifications, cost estimating, and cost analysis
14. Pavan A, Re Cecconi F, Maltese S, Oliveri E, Aracri G, Guaglianone MT (2014) La scheda prodotti interattiva di INNOVance. Costr Laterizio 155:60–63
15. Oliveri E, Aracri G, Guaglianone MT, Pavan A, Re Cecconi F, Maltese S (2014) La denominazione dei prodotti da costruzione in INNOVance. Costr Laterizio 156:58–62
16. Daniotti B, Pavan A, Re Cecconi F, Caffi V, Chiozzi M, Lupica Spagnolo S, Maltese S, Pasini D (2013) InnovANCE: La Banca Dati Italiana per la gestione del processo edilizio. InnovANCE: An Italian database to manage the building process. In: ISTEA conference
17. UNI 8690-3: 1984—Edilizia. Informazione tecnica. Articolazione ed ordine espositivo dei contenuti
18. UNI 9038:1987—Edilizia. Guida per la stesura di schede tecniche per prodotti e servizi
19. Bianchi L, Chiozzi M, Alessandro RD, Daniotti B, Di Fusco A, Galli M, Giorno C, Gulino R, Lupica Spagnolo S, Pasini D, Pavan A, Pola M, Rigone P (2014) L'ottimizzazione del processo edilizio attraverso una gestione efficiente delle informazioni. Building process optimization through an efficient data management. In: ISTEA conference
20. Lupica Spagnolo S, Pasini D, Pavan A, Daniotti B, Re Cecconi F, Oliveri E, D'Alessandro R, Di Fusco A, Pola M, Rigone P (2017) L'ottimizzazione del processo edilizio attraverso una gestione efficiente delle informazioni: classificazione, codifica e schede tecniche digitali. [The optimization of building process through an efficient management of information: classification, codification and digital datasheets] Edilstampa srl, Roma
21. Pavan A, Daniotti B, Re Cecconi F, Maltese S, Lupica Spagnolo S, Caffi V, Chiozzi M, Pasini D (2014) INNOVance: Italian BIM database for construction process management. In: ICCCBE conference
22. Caffi V, Re Cecconi F (2013) Oggetti BIM INNOVance per l'industria italiana delle costruzioni. INNOVance BIM Object for the Italian building industry. In: ISTEA conference
23. Re Cecconi F, Pavan A, Maltese S (2013) Gestione delle Informazioni nel Processo Edilizio. Information management in the building process. In: ISTEA conference

References

Chapter 3
Standardized Guidelines for the Creation of BIM Objects

Abstract The issue of BIM libraries and of standards for defining BIM objects useful for the information flow in the building process, is object of particular attention by the professionals of the building industry, also owing to various international research programs, among which of particular relevance there is the experience of the British National BIM Library under the umbrella of the NBS AAVV (BIM Object Standard V, 2.0, NBS, 2018 [1]). BIM technology, currently viewed by many as a driving force for the development and innovation of the building sector, requires operational standards useful to guarantee the sharing of information and especially information that it may be used by all professionals involved in the process AAVV (Level of development specification 2013, BIM Forum, 2013 [3]). The development of interoperable BIM libraries is a necessary condition for this to take place and it is a fundamental step to facilitate the dissemination of BIM in the entire process AAVV (Il processo edilizio supportato dal BIMM: l'approccio INNOVance, Collana INNOVance Edilstampa, 2017 [2]). The libraries of BIM objects, in fact, must guarantee quality and coherence in terms of geometry, related information content and expected behavior within the scope of a model and of procedural operations in general. In particular, the information content must be such to meet process requirements, which is not something obvious given the complexity of the sector. Such requirements are fundamental to allow professionals to be able to use BIM with no hesitation whatsoever when managing the entire process, starting from the design to the management of the actual constructed facility (AAVV (2012) Common BIM Requirements 2012, vols 1–12. COBIM V1.0 Finland [4]). This chapter illustrates the process for defining BIM libraries, whose information content is strictly based on the structure developed with the INNOVance database. The latter, owing to the exhaustiveness of the categories and technical datasheets used, offers an overarching and flexible support, capable of evolving depending on the information needs defined by the professionals of the sector. Therefore, the objects based on said database guarantee the abovementioned requirements of coherence and completeness. The indications provided below were drawn up during the realization of BIM objects supported by the INNOVance database, and concern the choice of the objects, the encoding for the database, the modeling, the verification of coherence, of information completeness and of interoperability. Moreover, indications have been provided on the contents and modalities for the graphic presentation of the objects.

© Springer Nature Switzerland AG 2020
B. Daniotti et al., *BIM-Based Collaborative Building Process Management*, Springer Tracts in Civil Engineering,
https://doi.org/10.1007/978-3-030-32889-4_3

3.1 Definition of BIM Objects

The development of BIM objects and of the procedures for implementing an inter-operable data modeling and sharing—illustrated below—is the result of a complex work dedicated to the production and experimentation of BIM libraries, with the involvement of Politecnico di Milano, Politecnico di Torino and One Team, software distributor and developer [6]. The objective of the work was to establish criteria for the development and production of BIM libraries, whose objects were to be provided with information content based on the structure of the INNOVance database [8].

The software to coordinate and allow communication between the INNOVance database and the BIM solution chosen for the modeling was developed by One Team, which along with Politecnico di Torino also drew up the texts relating to the information content of the objects.

Also Consorzio TRE was involved in the development and test phase of the BIM system.

Politecnico di Milano, besides producing a series of BIM objects, coordinated the activity.

The procedure for processing BIM objects was managed in four subsequent phases (Fig. 3.1):

1. analysis and definition of the building's object models and technical solutions;
2. encoding according to the INNOVance standard;
3. modeling in BIM environment and association of the INNOVance code;
4. organization of the objects for the due verifications of interoperability according to buildingSmart standards.

The first modeled objects, whose criteria of choice are illustrated further on, were drawn from Assimpredil ANCE's repertoire [5].

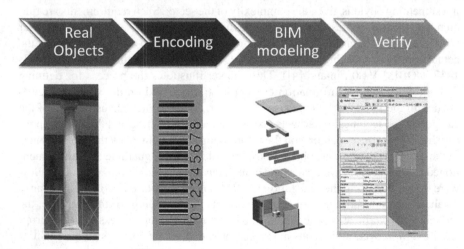

Fig. 3.1 The workflow for the preparation of the libraries of BIM objects

The buildings of reference for the modeling actions were the INNOVance's model of reference, processed by Politecnico di Milano, as well as real case studies: Politecnico di Torino modeled the components of a building of the University campus, and a group of Politecnico di Milano took care of modeling some facilities of the Libeskind residence within the scope of the intervention CityLife a Milano.

The modeling of BIM objects was conducted in Autodesk's Revit 2013 and following versions up to 2015. Autodesk participated as technical partner of the INNOVance project through its distributor One Team.

The work of INNOVance did not take into consideration other BIM software, for two fundamental reasons.

First of all, different BIM solutions use different strategies for defining different libraries; to this regard, see Table 3.1. Moreover, not all the attributes of the objects contained in the INNOVance database can or must be embedded in the objects modeled with a BIM.

For this reason, it was preferred to develop the issue of the interface between the INNOVance database and a single BIM software, with an approach that can be repeated also with other software, despite being proprietary software.

The INNOVance code associated to the products of the database must be associated to each BIM object: in fact, the code guarantees access to the properties contained in the database.

Exporting the object, with the code associated to it, in the interoperable format IFC of buildingSmart International, guarantees the interoperability of the INNOVance library with any other certified software in accordance with buildingSmart's standards.

One Team took care of developing the interface between the BIM modeler and the database and related procedures: to associate the objects to the INNOVance code; to verify the congruence between the models and the corresponding attributes of the database (Fig. 3.2).

Therefore, a brief description is provided below of the characteristics of the component defined add-in INNOVance, which is the part of the system that allows to associate the INNOVance code to the objects and to conduct the necessary verifications.

The add-in allows to manage the association of the encoding of the central system to all project objects:

– project file
– objects that identify functional areas and destinations
– architectural objects
– structural objects
– plant-engineering objects
– others.

The objects, with the materials and components of which they are composed, were first encoded and registered in the database and then modeled and associated to the code using the INNOVance add-in. Subsequently, the interoperability was

Table 3.1 Compared analysis of the main characteristics of BIM objects in the most widespread authoring tools

Software	Terrain	Wall	Column	Slab	Beam	Roof	Stairs	Doors	Windows	Curtain wall	Ceilings	Libraries
Allplan nemetschek	Existing	Structural and not structural It allows to define multiple and different layers	Existing	Existing	Existing	Existing	Existing	Existing	Existing	Existing	Existing	Smartparts
Archicad graphisoft	Existing	Structural and not structural It allows to define multiple and different layers	Existing	It is allowed to define multiple and differente layers	Existing	Existing	Existing	Existing	Existing	Existing	Existing	GDL

(continued)

Table 3.1 (continued)

Software	Terrain	Wall	Column	Slab	Beam	Roof	Stairs	Doors	Windows	Curtain wall	Ceilings	Libraries
Architecture bentley	Existing	Structural and not structural It allows to define multiple and different layers	Existing	Existing	Existing	Existing	Existing	Existing	Existing	Existing	Existing	Parametric cells
Revit autodesk	Existing	Structural and not structural It allows to define multiple and different layers	It is allowed to define structural columns and architectural columns (not bearing loads)	It is allowed to define multiple and differente layers	Existing	Existing	Existing	Existing	Existing	Existing	Existing	Families of components

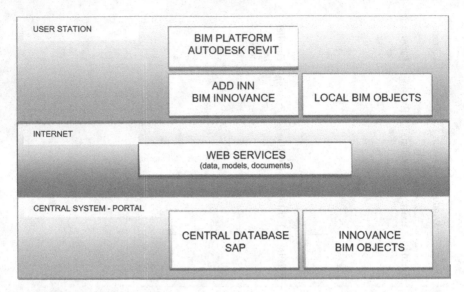

Fig. 3.2 Scheme of INNOVance BIM subsystems

verified to evaluate how the defined flow can guarantee the integrity and sharing of the information contained in the INNOVance database.

The BIM primitives in RVT format were made available in the format buildingSmart IFC 2X3 and they were controlled again with the software Solibri Model Viewer v9.0, to verify the congruence of the files in the IFC 2X3 format with Revit objects. This allowed to verify that the attributes of the RVT format are readable also in the IFC format, in particular the univocal code that allows to associate each object to the INNOVance database.

The strategy used by the project provides, in fact, for the BIM primitives to be associated to a univocal code, owing to which it is possible to access the data contained in an external technical datasheet in text format, which is thus open and interoperable.

This allows all information necessary in any phase of the building process to be made available by the technical datasheet associated to the BIM primitive.

The cooperation with the producers' associations allowed the technical datasheet developed by INNOVance to contain both general and specific data relating to the producer.

This way of operating makes the maintenance of the BIM primitives simpler: when, for any reason, it is necessary to update the data of the datasheet of a product or of a component, it is possible to intervene exclusively on the datasheet and on the records of the database, without having to redefine the relevant BIM objects.

In fact, the univocal code binds the objects to the datasheet. Therefore, each updating of the data contained in the relevant datasheet is accepted by the primitive BIM. In this way, BIM objects do not need to be remodeled and the maintenance of the BIM library is a lot easier.

The general structure of the work and the successful verifications of interoperability led to the conclusion according to which the established workflow and the characteristics of the INNOVance libraries meet all the requirements that authoritative international sources [9] consider to be the index of an advanced solution, if not ideal, for the development of a BIM library:

– General and specific information relating to the producer;
– Users, based on the entire industrial process (building life cycle);
– Classification of objects, based on unified terms and on their properties;
– Objects provided in BIM format with the full support for IFC interoperability.

3.2 Definition of Models of Reference and Choice of Technical Solutions

The model defined by the INNOVance elementary model of reference was designed to provide a generic building object, structured in such a way to include the most diverse cases of technical solutions, that are not necessarily available in a real case study: closings elements on the ground, on open spaces, horizontal and inclined closings, facilities and other technical elements. The model, represented in Fig. 3.3, is a building made of a basement and two floors above the ground.

The technical solutions processed for such model were drawn from Assimpredil ANCE's repertoire [5].

The repertoire provides a wide collection of building models, of which an example has been provided in Figs. 3.4 and 3.5.

The BIM objects developed are ascribable to the following sub-system categories:

– horizontal and vertical closings

   a. walls
   b. coverings
      i. horizontal
      ii. inclined
   c. floors
      i. on the ground
      ii. on open spaces
   d. external doors

– vertical and horizontal divisions

   a. walls
   b. internal floors
   c. banisters
   d. internal floors

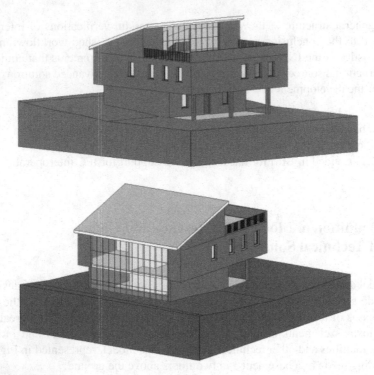

Fig. 3.3 The INNOVance elementary model of reference

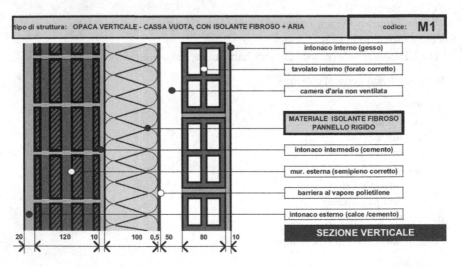

Fig. 3.4 Example of a wall object model from ANCE's repertoire

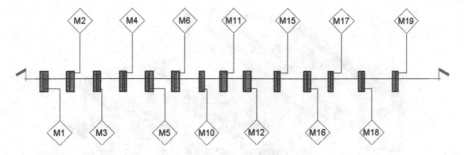

Fig. 3.5 BIM INNOVance objects: catalogue of walls according to assimpredil's repertoire of technical solutions

– connection elements

  a. stairs
  b. structures
     i. of foundation
     ii. of elevation

– structural elements

  a. foundation structures
  b. elevation structures

– systems

  a. thermal system
  b. water distribution system.

Before defining the modeling, the minimum energy requirements of the technical solutions to be chosen were defined.

In accordance with the Decree of the President of the Republic no. 59/2009, the following restrictions were established:

– vertical opaque structures: $U < 0.62$ W/m^2 K;
– coverings: $U < 0.38$ W/m^2 K;
– floors towards not heated environments or toward external environments: $U < 0.65$ W/m^2 K.
– doors: $U < 4.6$ W/m^2 K.

The Reference Model was useful for various types of simulations and analyses:

from those relating to the levels of implementation of the model, to different types of modeling such as, for example, the structural modeling (Fig. 3.6) and analyses such as the quantity take off.

From a structural viewpoint, the entire building is realised in reinforced concrete.

The structures of vertical elevation are identified in the basement by full perimeter walls on the ground and by pillars in the central parts. The development above the ground, instead, is realized entirely through pillars. It is important to highlight the

Fig. 3.6 Three-dimensional overall view of the structure of the INNOVance elementary model of reference

particular configuration of the second floor above the ground, where a sharp rise was created from which the offset pillars branch off.

The horizontal structures are supported by principal and secondary reinforced concrete beams and, depending on the floor of reference, they can be in concrete and masonry or in predalles. Also, the inclined covering is realized through the same structural technology.

In the centre of the building, the design provides for a stairwell with reinforced concrete separators that start from the basement and arrive up to the inclined covering, thus creating an effective bracing nucleus.

The foundations are continuous along the whole external perimeter, while a small bed supports the walls that constitute the stairwell. In the central part supporting the pillars there is a foundation grid that connects to the continuous perimetric area.

On the basis of the evaluations emerged from the activities conducted, the primary objective of the experimentation mainly concerned the verification of information congruity between the encoding system proposed by INNOVance (generated in SAP with the purpose to define a univocal alphanumeric code associable to the different technological systems) and the relevant association to the components—building and not—present in the parametric models processed according to the regulatory provisions on Public Works.

3.3 The Instruments: Revit and the INNOVance add-in

The BIM platform selected for the activities with the aim to develop the INNOVance project was Autodesk Revit. The BIM Revit software is developed specifically for Building Information Modeling (BIM) and allows designers and builders to face all phases of the project, from the conceptual phase to the building phase, with a coordinated and homogeneous approach based on the model.

Autodesk Revit includes functionalities for the architectural design, for MEP engineering, for MEP engineering and structural planning and for construction. Moreover, Revit has all the instruments necessary for development. In particular, the APIs provided with the software allow to extend the data model creating "INNOVance" objects with dedicated properties and integrated managerial data, to extend the user's interface in order to create new commands and user forms that can interact both with the objects of the model and with the underlying database. Such forms also allow to widen the functions of connectivity and interface with the external world, in particular with the web services exported and provided by the INNOVance Portal.

The architecture of the communication system between the BIM design software and the Portal was planned following SOA standards (Service Oriented Architecture) that provide for an architecture based on web services. The portal becomes an agent of the distributed system that provides different types of services (interrogation, encoding, etc.) towards satellite systems (for example, BIM design software).

Specifically, the INNOVance architecture was structured with the following characteristics:

Access to the INNOVance Portal through the development of webservices with methods and functionalities dedicated to the exchange of information and of functions with the designers' positions.

Additional functionalities for the BIM design software through the development of software for the communication and use of the portal's webservices (webservice consumer) and the webservice client that uses the portal's standard methods and functionalities for the communication and exchange of information in XML open format.

The summary scheme of how SOA technologies are applied to the BIM INNOVance is presented in Fig. 3.7.

The application model uses the concept of maximum independence of the local system (Client: the user's design workstation) from the central system. This implies that the central system has to allow the interrogation not only of the data contained in it, but also and especially of the data structure; only in this way, the client system (the designer's software) remains always independent from the hardware and from the configuration of the device on which it is installed. The architecture of the BIM INNOVance system was designed exploiting the most modern distributed system concepts based on web services architecture. Such requirements originate from the need to develop a system capable of being utilized by remote users on their positions. The possible configurations of the single users are many. Therefore, it is not possible to plan a priori the single individuals' configurations or limitations. In this view, the

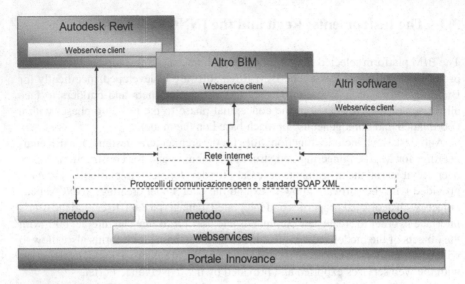

Fig. 3.7 Relational scheme of INNOVance subsystems

user's local system configured as Workstation with the BIM design platform installed (in the specific case Autodesk Revit) must provide for the installation of the software components of the BIM INNOVance system dedicated to extend the user's interface with the new commands. Moreover, it must be able to communicate with the central system through the Internet, utilizing the services made available by the portal. It is necessary to avoid installing on the PC local platform complex databases and copies of the central INNOVance database. In fact, due to the nature of the database, it would become obsolete in a short timeframe and would require continuous alignments, in addition to more complex hardware and software requirements.

The INNOVance software must be installed on the user's BIM workstation. That means installing an add-in for Autodesk Revit. The add-in is visible within the BIM software as additional Ribbon-Bar "INNOVance." The new tools allow to interact with the BIM model exploiting the new extension functionalities of the data model and to verify the encoding data. Moreover, the add-in allows to manage automatically all the connections to the web services of the portal necessary for the correct functioning of the BIM INNOVance application. The user's hardware must be connected to Internet, in order to be able to use the web services of the central portal. The designer can anyway work disconnected in the phases relating to the design and creation of the BIM model. However, it is necessary to be connected for the functions that require an exchange of information and data with the Portal (e.g. research of materials, encoding of objects, etc.).

The INNOVance BIM system allows a free expansion of the central data model owing to the fact that the Client's system can recover all the information on the typology and quantity of the characteristics associated to an object, in order to use this information in a parametric way and in a totally expansible way. The information

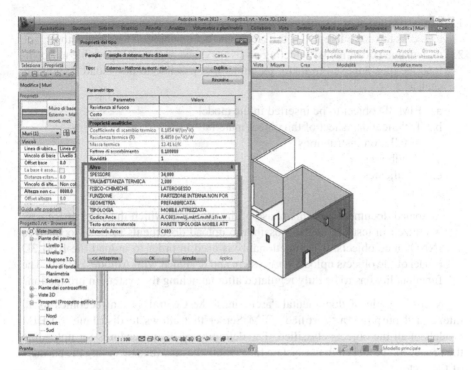

Fig. 3.8 Revit interface and additional INNOVance properties

associable to the BIM model from the INNOVance database is not limited to the sole encoding of the object. In fact, all the information characterizing the object, or the material is associated to, for example:

– Geometrical characteristics
– Typological characteristics
– Performance of functional characteristics
– Descriptive characteristics
– Procedural characteristics
– Etc.

Figure 3.8 shows the result of an association of a BIM object with an Object of the Central system; the properties visualized are all characteristics that to date are provided for the INNOVance object selected.

Moreover, the BIM INNOVance system allows to associate and archive documents. In particular, with reference to the BIM design methodologies, the system is capable of archiving the technical documentation referred to the objects contained in the model. The documental System exploits the archiving technologies of the SAP platform and it allows to memorize any type of electronic file. Specifically, it is possible:

1. to consult the list of documents: starting from an INNOVance object, it is possible
   to access the complete list of the electronic documentation associated to it
2. to download documents: selecting an INNOVance object, it is possible to download the files annexed to it; for example, it is possible to download documents
   such as:

   a. BIM 3D object to be inserted in the model
   b. Technical datasheet of the object/material
   c. Installation instructions
   d. Handbooks
   e. Certificates
   f. Etc.

3. to upload documents: in this phase of the project in which the BIM instrument
   is utilized in tests, it is also possible to upload documents to be associated to
   INNOVance objects. The functionality is used in order to upload the BIM 3D
   model of the objects uploaded to system and designed with Autodesk Revit. The
   functionality has to be duly regulated after launching the system in production.

With this logic of documental "archiving," the Central system becomes to all
intents and purposes a "certified" BIM Server that allows to distribute 3D BIM
models of all the objects classified. The definition of the procedures for uploading
the producers' models into the system falls within the activities involving the partners
of the sector.

3.4   Encoding According to the INNOVance Standard

A univocal INNOVance code was assigned to each object to be modeled, with the
SAP codifier, which is the software on which the INNOVance database is based.

SAP and the encoding procedure are not covered in this chapter because they do
not concern the modeling strictly speaking. However, it is important to mention that
the INNOVance code generated by SAP is composed of seven fields that correspond
to just as many main characteristics of families of technical elements, and that they
were identified in the previous research phase dedicated to the univocal encoding
of the construction products and components. Moreover, SAP generates a univocal
identification for each object, the so-called ID SAP, as well as a brief description of
the main characteristics of the object, defined also INNOVance Name.

The encoding procedure in SAP allows to associate, to each construction component to be modeled, a basic and detailed bill—BOM—of the materials and components that constitute the object, in turn encoded and registered in the INNOVance
database.

This allows each object to be identified in all its characteristics, owing to the
INNOVance code and to the ID SAP.

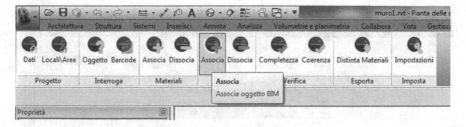

Fig. 3.9 Add-in INNOVance tools

This guarantees that all the attributes of a component are registered in the database, are accessible to all because in text format, and are available for the user. The system allows the encoding of any object inserted in the project.

In order to research and select an INNOVance object in the Central system, the designer selects the add-in command in the Revit interface (Fig. 3.9); the system opens the research and interrogation window that communicates via the Internet with the web services of the Central system. The research can take place setting the type of system, work, typology and material.

After selecting the type of object, all the characteristics present in the Central system (read dynamically) are shown in the left part of the window, allowing to apply filters on all properties of the objects/materials. Selecting the button "Apply research filters …" the central system is interrogated, and the identified data are shown in the right part of the window (Fig. 3.10).

The system can create new properties in the objects and exten the data already inside the model. It always verifies the congruence of the data with the dimensional parameters of the INNOVance database. Moreover, it warns the user of any anomalies

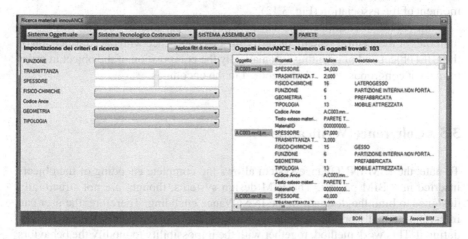

Fig. 3.10 INNOVance interface for accessing the database and for the connection with the BIM objects in Revit

Fig. 3.11 INNOVance instruments for the verification of stratigraphies in revit

and of the geometrical and informative congruence of the model. With regard to elements composed of strata (Walls, Coverings, Slabs and False Ceilings), there is a further control on the congruence between the elements in Revit and in the INNOVance bill of materials and on the association of the various characteristics with the materials composing the strata. Also, in this case the incongruence is highlighted, and the user is requested to input any missing information concerning the layers of which the selected Revit object is composed (Fig. 3.11). The result of the operation is the automatic creation of BIM properties and their enhancement on the basis of the INNOVance code selected and on the basis of the characteristics available at the moment of the association (Fig. 3.12).

With regard to stratified objects, the system is capable of reading the bill of materials of the Central system, of verifying the congruence with the stratification of the BIM object and of updating automatically the composition of the object in order to make it congruous with what specified in the encodings.

3.5 Coherence Verification

To date, the BIM INNOVance system allows the complete encoding of the objects inserted in a BIM project. The BIM design systems, though, are not structurally designed to bind the designer to the INNOVance encoding. Therefore, the user can intervene manually on the information downloaded from the Central system, invalidating it. This work method, together with the impossibility to modify the behaviors of the BIM design software, obliged—from the very outset—to structure control and validation functionalities on the INNOVance datum contained in the BIM project.

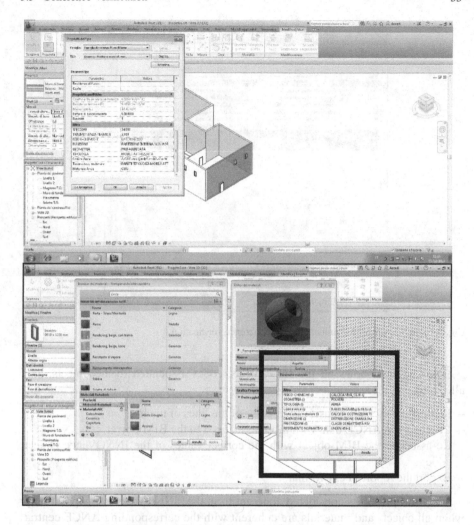

Fig. 3.12 Properties of associated material through the INNOVance add-in

This especially in view of a transfer of project information to the Central system. The verification of the information is conducted by controlling the INNOVance code associated to the objects. The system is capable of generating a completeness report and to indicate the dissimilarities between the information saved on the designer's local model and the values present in the INNOVance central database. The Coherence control (Fig. 3.13) allows to verify if all the values of the properties, for the objects encoded ANCE, are compliant with what present in the Central system (that is, that INNOVance parameters have not been modified manually).

Fig. 3.13 INNOVance instruments for the coherence verification

The interface shows all the INNOVance codes associated to the objects, and allows to:

– Identify errors on the basis of the color of the lines of the list
– Compare the ANCE data with those associated to the BIM object
– Re-associate the BIM object with the correct values of the INNOVance Object code
– Export the report in csv format.

The BIM project is considered coherent from the "INNOVance" viewpoint only when all objects and materials are coherent with the corresponding ANCE central code.

The command opens the mask of the bill of materials grouped per code providing quantity data, possible area and volume and, for the elements composed of strata (Walls, Coverings, Slabs and False Ceilings), the calculation of the material composing each layer (Fig. 3.14). Moreover, the command allows to extract the bill of materials as CSV and XML formats to be used for the future uploading on the INNOVance portal.

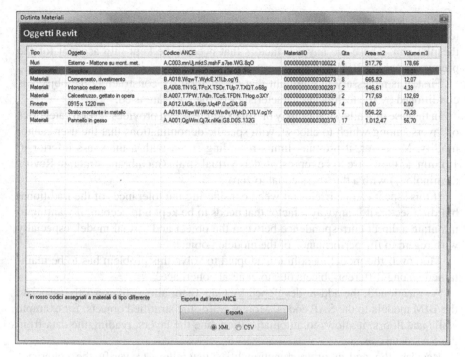

Fig. 3.14 INNOVance instruments for the production of bills of materials

3.6 Modeling of BIM Objects

The originality of the INNOVance approach to BIM modeling—starting from the definition of primitive BIMs (or BIM objects)—is that it has recognised the impossibility to associate objects, in the modeling software, with all the attributes and characteristics of the real objects classified in the database, and needed during the building process.

The solution adopted is to associate a univocal code to the BIM primitives, which recalls an external, open and interoperable technical datasheet, in text format.

The technical datasheet associated to the BIM primitive with a code, makes all the necessary information available, in any phase of the building process.

Moreover, should the datasheet of a product or of a component be updated or modified in order to add further characteristics not yet taken into consideration, this only entails an intervention on the datasheet and on the database records. In fact, it is not necessary to redefine the related BIM objects, that anyway remain bound to the datasheet owing to the univocal code, thus receiving as obvious consequence every updating of the data contained in the datasheet.

In this way, the BIM objects do not need to be remodeled and the maintenance of the BIM library is a lot easier.

When developing the BIM libraries, critical aspects were identified that open up to just as many possibilities for the development of following implementations of the platform. These are anyway factors that need to be kept into account for the preparation of the BIM objects.

First, it is necessary to highlight that the method for constructing the objects is conditioned by the technical characteristics of the software used.

In the case of walls' stratigraphy of, for example, Revit provides several categories of layers among which to choose, with specific denominations that the user cannot modify. Moreover, it presents limits regarding strata with a thickness inferior to 1.6 mm, as they have to be represented as virtual strata (membrane strata, in Revit's terminology) with a thickness equal to zero.

If this aspect can be irrelevant when considering the tolerances of the traditional building sector, it is anyway a factor that needs to be kept into account if wanting to maintain a direct correspondence between the object and virtual model, especially with regard to the performance of the modeled object.

Moreover, the modeling solution adopted to solve this problem has to be maintained among different objects due to general coherence.

As mentioned, the add-in developed by One Team allows to correctly associate the BIM models to the SAP codes. Moreover, for the stratified objects, for example walls and floors, it allows to automatically define the layers, reading the data from the SAP database.

Besides, the add-in offers functionalities that allow to verify the coherence between a BIM object associated to a specific INNOVance code, and the real characteristics contained in the database:

For example, if the code of a window is associated to a wall object, the system can identify the error. If the example refers to a gross mistake, the system is capable of carrying out verifications at more refined levels, such as those of the stratigraphy of a wall.

Nonetheless, the add-in still does not allow the automatic mapping of the INNO-Vance materials with the Revit materials: as a consequence, the BOM (bills of materials) must be anyway defined with SAP's support, unless a manual mapping is carried out with the Revit materials database.

From SAP's viewpoint, the INNOVance encoder can associate different bills of materials to the same type of technical element (for example to make a distinction between different producers).

The add-in INNOVance for Revit instead is not capable of implementing different bills of materials associated to a same code.

For this reason, the preferred solution in this case was to make the correspondence in the database between code and bill of materials univocal.

Another critical aspect of add-in INNOVance, overcome however by the different implementations of the project, consists in the fact that it is directly connected to the specific Revit release.

The latter is an instrument in constant evolution and updating: currently Autodesk releases at least one new version every year.

This entails the need to update the add-in periodically for new versions of the modeling software; One Team took care of implementing procedures that allow an easy upgrade of the software for different Revit releases. It is a very important characteristic, because it guarantees the continuity of the work in INNOVance and the possibility to constantly keep step with the evolution of the BIM authoring software chosen.

As mentioned, the current workflow begins with the encoding and definition of the objects in the database, with their relevant characteristics and attributes. The modeling phase in the BIM environment and association phase of the code takes place in a second moment.

The opposite process, starting from the modeling in the BIM environment, is one of the potential developments of the research: the BIM objects, that are anyway provided with a series of indispensable attributes for the modeling, could be used as source for the generation of a first encoding. Certainly, the attributes provided in the database should anyway be completed, since the specific attributes of the BIM software are not exhaustive with regard to the complexity of a real object. However, this path could be an alternative strategy to populate the INNOVance database. Moreover, during the production of the BIM objects, the need to define common modeling standards emerged.

First, the modeling templates have to be made available in the same format, containing graphic formats and the general structure of the modeling environment.

Besides, it is necessary to deal with the issue of the level of detail—LOD—of the modeling, due to the needs of the different scales throughout the design and representation phases.

3.7 Interoperability Verifications

The phase following the modeling consisted in verifying interoperability in IFC format [7].

The operation itself is elementary, because it consists in a simple exportation from Revit in the format buildingSmart IFC 2X3. With the software Solibri Model Viewer v9, the congruence of the files in the IFC 2X3 format with the original ones was verified.

The attributes of the format RVT2013 resulted to be readable also in the IFC format (Figs. 3.15 and 3.16).

In particular, the univocal code and the ID that allows to associate each object to the INNOVance database is maintained when passing from Revit to IFC. This allows any IFC interoperable application to obtain information from the INNOVance database owing to the univocal code associated to objects. Interoperability verifications were carried out with success also on the entire construction model; in particular, on the structural model that was used to analyze the IFC quantity take off. The model exported in the IFC 2X3 format was analyzed with success in the Solibri environment.

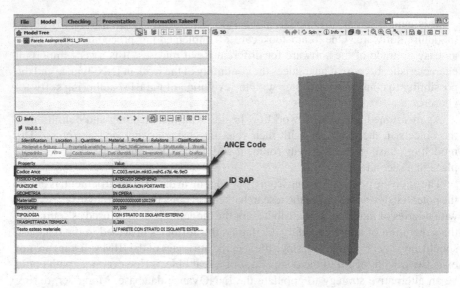

Fig. 3.15 Model of wall visualized in Solibri: the INNOVance codes highlighted are those that are returned whole from the model in the IFC format

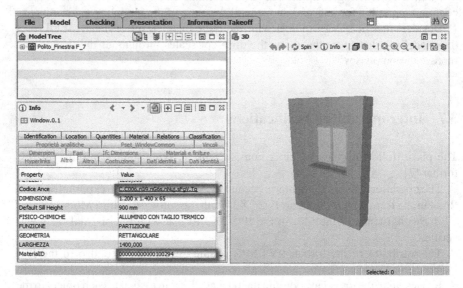

Fig. 3.16 Model of window visualized in Solibri: the INNOVance codes highlighted are those that are returned whole from the model in the IFC format

Figure 3.17 shows the QTO; Fig. 3.18 highlights the configuration of the structures and of the internal bracings, and the connection to the database.

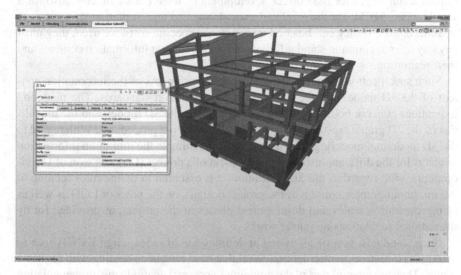

Fig. 3.17 Quantity take off of the structural elements carried out on the IFC file

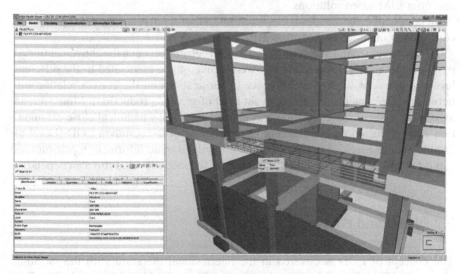

Fig. 3.18 Interoperability verifications of the IFC models configurating the reinforced structural elements of the INNOVance elementary model of reference

3.8 BIM and Modeling and Representation Standards

Fundamental issues for the correct development of INNOVance objects, also for a suitable use of BIM, are those related to modeling and representation strategies.

Indeed, even if they can be conditioned by the specific software used, they must anyway refer to common standards that guarantee coherent information contents and their restitution.

Such specifications concern the minimum requirements of the information content of the 3D model and the 2D graphic representation standards. The modeling indications concern both the information content of the 3D model, and the modalities of the 2D graphic restitution, as well as the modeling procedures. With regard to the 3D modeling, modeling rules are necessary relating to the minimum information contents for the different levels of development of a project (LOD—Level of Development). With regard to the 2D restitution, it is essential to define rules relating to the minimum graphic content of the project designs, on the basis of LOD as well as of representation scales and development phases of the project, as provided for by the national regulations on public works.

Such standards, first of all aimed at defining the libraries, align INNOVance to analogous initiatives at international level that have long been dealing with the same issues. They are integral part of the guidelines necessary to guide the implementation of BIM. Moreover, they constitute a tool aimed at controlling and verifying the quality of the BIM models and at evaluating the processes and competences of companies that offer BIM based solutions.

In perspective, the standards can constitute a useful reference for the training and education of qualified and certified BIM operators.

After various technical meetings and fruitful debates between the abovementioned partners, the main graphic standards were defined to be associated to the different levels of detail (with explicit reference to verifications on the correct relationship between scale/contents and the relevant level of graphic production). Moreover, an associable scale of representation was defined, and a new technical close examination was conducted on the different interpretations of the acronym LOD (Level of Development/Level of Detail), to which not only graphic information patrimonies can refer, and on which the different grades of the model's reliability can depend.

3.9 LOD and Grade

The concept of LOD is covered in the international literature with the meaning of Level of Development of an element or of a group (AAVV 2013-2 [3]). This interpretation, according to the definition provided by the BIM Forum corresponds to the grade of reliability that a project group can obtain from the data contained in a model.

It is important to keep into account that the concept of LOD does not correspond to the concept of scale, nor to the concept of graphic representation of an element: a poor representation in graphic terms can be very rich in terms of information content associated to an element in the database; vice versa, a very detailed graphic representation can lack the information attributes directly connected to the model.

For this reason, with regard to the graphic representation of the elements, the concept of Grade was introduced—analogously to concepts expressed in the international literature—to express the level of graphic detail required by the 2D representation on the basis of the different LODs and of the type of representation required on the basis of the different phases of in depth study of an intervention.

All this was put into relation with the Italian regulations on public works, such to contextualize the concepts already expressed with regard to the needs of the national building industry.

The INNOVance BIM objects interpret the concept of LOD according to eight different levels:

- LOD 000 Corresponds to level NM (Not Modeled) present in the AIA regulations. They are elements inserted in the model, but lacking the uses allowed. The objects can be modeled also in an accurate way, but they cannot be used for other project designs.
- LOD 100 The contents and uses allowed are drawn from the definitions of the BIM Forum. The level corresponds to that of the feasibility study. An example of LOD 100 is that of a building modeled from a planovolumetric viewpoint, without any detail whatsoever.
- LOD 200 The contents and uses allowed are drawn from the definitions of the BIM Forum. The level corresponds to that of the preliminary project.
- An example of LOD 200 is a planimetry defined from a dimensional and distributive viewpoint, but not with regard to construction choices.
- LOD 300 The contents and uses allowed are drawn from the definitions of the BIM Forum. The level corresponds to the definitive project. An example of LOD 300 is a planimetry showing in full the layers of the technical elements, that therefore provides precise indications concerning the construction choices.
- LOD 350 The contents and uses allowed are drawn from the definitions of the BIM Forum. The level corresponds to that of the executive project.
- LOD 400 The contents and uses allowed are drawn from the definitions of the BIM Forum. The level corresponds to that of the construction project.
- LOD 500 The contents and uses allowed are drawn from the definitions of the BIM Forum. The level corresponds to that of the As Built.
- LOD 550 This level contains all the details of LOD 500. It is utilized for the management and maintenance of the work.

It is useful to highlight that the concept of LOD, as interpreted by INNOVance, does not refer directly to the single objects of the model of a facility: therefore, the same model can contain objects with different LODs.

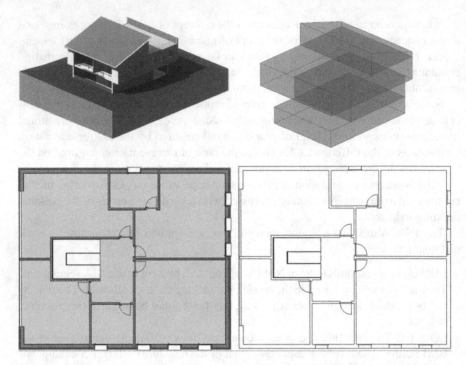

Fig. 3.19 Clockwise starting from above on the left: LOD 000, LOD 100, LOD 200, LOD 300

Figure 3.19 shows several LODs on the basis of the INNOVance elementary model of reference:

LOD 000 is a rendered sketch, that is not associated to any use allowed and that therefore does not provide information.

LOD 100 gives a general dimensioning of the volumetries; other data, such as surfaces, can be easily inferred.

LOD 200 provides a more precise indication on dimensions and internal layout.

LOD 300 defines with a certain precision the construction typologies of shell and divisions.

INNOVance decided to introduce the concept of Grade—found in the AEC (UK) BIM Protocol V2.0—with the meaning of Level of accuracy. The aim was to define the minimum grade of graphic and modeling accuracy required depending on the different LODs and of the necessary representation for the different phases of an intervention.

INNOVance's Grades are defined according to the following list:

– Grade 0 (G0)—Schematic design—Symbolic designs, placeholder, that can be out of scale, or can have no specific dimensions. This is the case of the electric symbology, that can lack a corresponding 3D. The minimum dimensions of the objects should not be inferior to 10 or 5 cm.

Table 3.2 Evaluation matrix LOD-grade INNOVance

Elementi tecnici	Studio di fattibilità/concept			Progetto preliminare			Progetto definitivo		
	LOD	Grade	Note	LOD	Grade	Note	LOD	Grade	Note
Struttura di fondazione	100	0		100	1		300	3	
Struttura di elevazione	100	0		100	1		300	3	
Chiusure verticali opache	100	1	disegno della massa concettuale	200	2		300	4	
Chiusure verticali trasparenti (facciata continua)	100	0/1	disegno della massa concettuale	200	2		300	3	
Chiusura orizzontale inferiore solaio a terra e su spazi aperti (escluse struttura)	100	0		100	2		300	4	
Chiusura superiore copertura continua o discontinua (esclusa struttura)	100	1	disegno della massa concettuale	200	2		300	4	
Chiusura superiore/infissi esterni orizzontali	100	0		100	2		300	3	
Partizioni interne verticali	100	0		200	2		300	3	

(continued)

Table 3.2 (continued)

Elementi tecnici	Studio di fattibilità/concept			Progetto preliminare			Progetto definitivo		
	LOD	Grade	Note	LOD	Grade	Note	LOD	Grade	Note
Infissi esterni verticali	100	0/1		100	2		200	2	
Infissi interni verticali	100	0		100	2		200	2	
Partizioni interne orizzontali/solai	100	0		200	2		350	3	
Partizioni interne inclinate/scale interne	100	0		200	2		300	3	
parapetto	000	0		100	1		200	2	
…									
Impianto di climatizzazione	100	0		200	0		300	1	3D di massima
Impianto idrico sanitario	100	0		200	0		300	1/2	
Impianto di smaltimento liquidi	100	0		200	0		300	1	3D di massima
Impianto elettrico	100	0		200	0		300	0/1	
…									
Arredi				100	1	richiesto solo layout arredi	100	2	

(continued)

Table 3.2 (continued)

Elementi tecnici	Studio di fattibilità/concept			Progetto preliminare			Progetto definitivo		
	LOD	Grade	Note	LOD	Grade	Note	LOD	Grade	Note
Apparecchi illuminanti				100	0		200	1	

Elementi tecnici	Progetto esecutivo			Progetto costruttivo			As built		
	LOD	Grade	Note	LOD	Grade	Note	LOD	Grade	Note
Struttura di fondazione	350	4		400	4	allegate schede tecniche dei materiali in appalto	400	4	allegate schede tecniche dei materiali utilizzati
Struttura di elevazione	350	4		400	4	allegate schede tecniche dei materiali in appalto	400	4	allegate schede tecniche dei materiali utilizzati
Chiusure verticali opache	350	4	allegati dettagli in 2D	400	4	allegate schede tecniche dei materiali in appalto	500	4	allegate schede tecniche dei materiali utilizzati
Chiusure verticali trasparenti (facciata continua)	350	3	allegate schede tecniche della tipologia del modello selezionato, con allegati dettagli in 2D	400	3	allegate schede tecniche con specifica di marca e modello	500	3	allegate schede tecniche dei materiali utilizzati
Chiusura orizzontale inferiore solaio a terra e su spazi aperti (escluse struttura)	350	4	allegati dettagli in 2D	400	4	allegate schede tecniche dei materiali in appalto	400	4	allegate schede tecniche dei materiali utilizzati

(continued)

Table 3.2 (continued)

Elementi tecnici	Progetto esecutivo			Progetto costruttivo			As built		
	LOD	Grade	Note	LOD	Grade	Note	LOD	Grade	Note
Chiusura superiore copertura continua o discontinua (esclusa struttura)	350	4	allegati dettagli in 2D	400	4	allegate schede tecniche dei materiali in appalto	500	4	allegate schede tecniche dei materiali utilizzati
Chiusura superiore/infissi esterni orizzontali	350	4		400	4	allegate schede tecniche dei materiali in appalto	500	4	allegate schede tecniche dei materiali utilizzati
Partizioni interne verticali	350	3	allegati dettagli in 2D	350	4	allegate schede tecniche dei materiali in appalto	400	4	allegate schede tecniche dei materiali utilizzati
Infissi esterni verticali	350	3	allegati dettagli in 2D	400	4	allegate schede tecniche dei materiali in appalto	400	4	allegate schede tecniche dei materiali utilizzati
Infissi interni verticali	350	2	allegati dettagli in 2D	400	3	allegate schede tecniche dei materiali in appalto	400	3	allegate schede tecniche dei materiali utilizzati
Partizioni interne orizzontali/solai	350	4	allegati dettagli in 2D	350	4	allegate schede tecniche dei materiali in appalto	400	4	allegate schede tecniche dei materiali utilizzati
Partizioni interne inclinate/scale interne	350	3		350	3	allegate schede tecniche dei materiali in appalto	400	3	allegate schede tecniche dei materiali utilizzati
parapetto	350	2	allegati dettagli in 2D	400	3	allegate schede tecniche dei materiali in appalto	500	3	allegate schede tecniche dei materiali utilizzati

(continued)

Table 3.2 (continued)

Elementi tecnici	Progetto esecutivo			Progetto costruttivo			As built		
	LOD	Grade	Note	LOD	Grade	Note	LOD	Grade	Note
...									
Impianto di climatizzazione	350	3		400	3		500	3	
Impianto idrico sanitario	350	¾		400	3/4		500	3/4	
Impianto di smaltimento liquidi	350	3		400	3		400	3	
Impianto elettrico	350	3		400	3		400	3	
...									
Arredi	100	3		000	0	escluso dall'appalto			
Apparecchi illuminanti	350	3	selezione apparecchi illuminanti e distribuzione	400	0/1	aggiunte indicazioni di posa e schede tecniche			

– Grade 1 (G1)—Concept—3D Model with the minimum level of detail possible. Useful for representations in scale 1:100 or 1:200.
– Grade 2 (G2)—Definitive designs—3D Model with graphic details enough to identify the main characteristics of the object: type, shape, dimensions and materials. It can include 2D detailed designs. The minimum dimensions of the objects should not be inferior to 1 cm. Useful for representations up to scale 1:50. For the representation of objects up to 0.5 cm, it is suggested to develop 2D designs instead of 3D, printable in scale 1:20.
– Grade 3 (G3)—Execution/construction designs—3D Models carried out with Grade 2, complete of more detailed 2D designs, printable in scale 1:10, 1:5, 1:2, 1:1. Some parts can be realized in 3D, if necessary.

In order to make explicit the relation between LOD, Grade and project phases, INNOVance proposed a matrix that puts such concepts in relation with construction elements of the building system, based on the phases provided for by the regulations on public works.

In Table 3.2, LOD indicates the reliability of the information and its allowed use, while the Grade defines the requirements of the element's graphic and modeling accuracy.

If drawn up in contractual phase, the matrix makes explicit, for the different project phases, the client's requirements in terms of development and use of the BIM model in each phase of the construction process.

References

1. AAVV (2018) BIM object standard V2.0, NBS
2. AAVV (2017) Il processo edilizio supportato dal BIMM: l'approccio INNOVance, Collana INNOVance Edilstampa
3. AAVV (2013) Level of development specification 2013, BIM Forum
4. AAVV (2012) Common BIM requirements 2012, vols, 1–12. COBIM V1.0 Finland
5. Borghi R (2008) Efficienza energetica e requisiti acustici passivi degli edifici. Assimpredil ANCE, Milano
6. Caffi V, Daniotti B, Lo Turco M, Madeddu D, Muscogiuri M, Novello G, Pavan A, Pignataro M (2014) BIMM enabled construction processes: the INNOVance approach in Energia. Sostenibilità e Dematerializzazione operative, Convegno ISTeA, Bari
7. Caffi V Re, Cecconi F (2013) INNOVance BIM objects for the Italian building industry. Convegno ISTEA, Milano
8. Daniotti B, Pavan A, Re Cecconi F, Caffi V, Chiozzi M, Lupica Spagnolo S, Maltese S, Pasini D (2013) INNOVance: an Italian database to manage the building process. Convegno ISTEA, Milano
9. Palos S, Kiviniemi A, Kuusisto J (2014) Future perspectives on product data management in building information modeling. Constr Innov 14(1):52–68

Chapter 4
Collaborative Working in a BIM Environment (BIM Platform)

Abstract The paradigm of the fourth industrial revolution (Industry 4.0) involves data management and the interconnection between machines-objects-people and processes. The key words of the revolution underway are:

1. information in real environments: AR (Augmented Reality);
2. data management: Big Data and A.I (Artificial Intelligence);
3. digital collaboration;
4. intelligent objects: IoT (Internet of Things);
5. additive manufacturing: 3D printing.

Compared to the previous industrial revolutions (18th century: mechanical loom vs. iron and steel; 19th century: production line vs. reinforced concrete, 20th century: automation manufacturing vs. precast concrete), today buildings are fully involved in Industry 4.0, alongside all other industrial sectors and services sectors. Therefore, the new approach to production and the use of products involves objects, subjects and processes, integrated among each other through the generation of common information in continuous evolution. Such information needs to be collected (in a structured way), processed (precisely and statistically), redistributed (openly and transparently). With regard to the quality of the works and to the competitiveness of the sector, Regulation no. 305/2011 (European Commission, Regulation (EU) No 305/2011 of the European Parliament and of the Council, Official Journal of the European Union, Strasbourg, [5]) obliges all producers to declare the technical characteristics of construction products before their commercialization. Today such information on products must be made available in an open and standardized form (Reg. no. 1025/2012, par. 3), guaranteeing data transparency for the public and private sectors. In the building sector, the main means to convey information has historically been represented by designs and related documents. In the first technological transit, from the drafting machine to CAD (Computer Aided Design), not much changed, if not in a merely technological sense, more than in a procedural sense. The use of CAD (vector software), as in the case of the drafting machine, involves the management of vectors and texts for the representation of concepts. The interaction between technicians takes place through the use of analogical or digital means, thus "static", remote from the process.

© Springer Nature Switzerland AG 2020   71
B. Daniotti et al., *BIM-Based Collaborative Building Process
Management*, Springer Tracts in Civil Engineering,
https://doi.org/10.1007/978-3-030-32889-4_4

4.1   BIM and Object Oriented Programming

The introduction of BIM (Building Information Model/Management) in the building
sector and in the design of objects in architectural and engineering modeling software,
changed such scenario. The new paradigm, in fact, no longer concerns vectors, terms
and files but the management of geometrical and non-geometrical data (alphanumeric
and multimedia), through the involvement of the machine. The computer becomes the
"dynamic" instrument of interaction, capable of reprocessing information, supporting
the technical and non-technical subjects involved.

The transit from CAD (means of "*representation*" of terms and vectors; typical
tool of the design phase) to BIM (means of "*simulation*" of geometrical data and
information; incremental tool for each phase of the process, from strategy to operation), requires the indispensible creation of a specific digital environment where
all information generated over time can be collected and managed: a *Common Data
Environment*.

Initially, this task was entrusted directly to the 3D-parametric *Model*, produced by
the BIM authoring software (Revit, Allplan, Archicad, Microstation, Edificius etc.;
Graphic model: geometrical objects with related information).

This does not allow, though, to obtain much more, or something completely different, from the scheme of the traditional process, based on analogical/paper based
means of information (Fig. 4.1). The sole introduction of BIM graphic models (object
planning) in the field of design, compared to paper based or CAD files (vector planning) does not entail an actual involvement of the entire sector throughout all phases

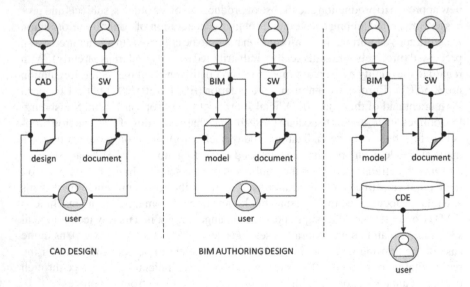

Fig. 4.1 Traditional CAD design; actual BIM authoring design; BIM (method) process

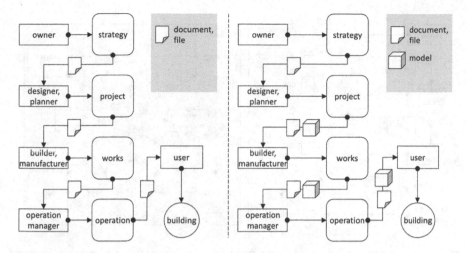

Fig. 4.2 Paper-based or digital CAD information process (on the left) and BIM digital information process, use of the

of the process: strategy, production and operation. The sole *Graphic Model* does not include and cannot contemplate the information of the entire sector that remains relegated to the traditional documents supporting graphic tables (Fig. 4.2).

4.2 Common Data Environment

In order to solve this critical aspect, the current consolidated need is to create, besides Graphic Models, specific Common Data Environments (CDE BS1192-1:2007, and PAS 1192-2:2013, UNI 11337:2017) supporting the entire information process (data collection, reprocessing and management). Besides being composed of Graphic Models (design), such information process also consists in a relevant number—if not higher—of additional documents, reports, calculations, acts, etc. supporting and integrating the mere design. This dataflow system—through the CDE—allows to involve all subjects (owner, designer and planner, manufacturer and builder, operation manager, user) and all phases of the building process (strategy, project, work, operation).

In these digital sharing environments (CDE) it is possible to carry out an integrated management of files, models (graphs) and documents relating to a project, to an intervention or an Asset, over time. Moreover, their use leads to stop using the traditional procedural flows, in an *antagonist* relationship (more CDEs differentiated for each phase of the process—see Fig. 4.3, on the left). Indeed, they spur to use *collaborative* procedural flows, supported by a single centralized CDE, on which all the actors involved operate. Such flows involve the entire lifecycle of the asset, from strategy to demolition and reuse of the land (see Fig. 4.3, on the right).

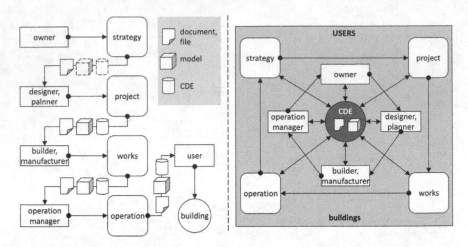

Fig. 4.3 The BIM information process with CDE's phases (on the left) and the BIM information process on a centralized CDE (on the right) sole Graphic Model (on the right)

4.3 CDE According to British Standards

This system was originally theorized by the British Standard (BS) in the voluntary regulation BS 1192-1:2007,[1] and was called *Common Data Environment* (CDE).

The scheme of the CDE regulated in BS 1192-1:2007 makes specific reference to the development phase of the building process (development, capex: strategy, design, construction) as the interaction between the various specialist design disciplines and the project leader and between the project/building teams, in their whole, and clients.

As represented in Fig. 4.4, anti-clockwise.

In the subsequent Technical Specifications of the British Standard PAS 1192-2:2013,[2] the use of CDE[3] [9] is specifically defined in a digital sense and expressly extended to every phase of the process, including the management and maintenance phases: execution, opex (operation and maintenance; execution phase, better defined in the following part 3 of the regulation, PAS 1192-3:2014) [10] (Fig. 4.5).

---

[1]BS 1192-1:2007—Collaborative production of architectural, engineering and construction information—Code of practice.

[2]CDE rules are those indicated in Regulation BS 1192-1:2007, since the sole additional specifications of the specific technical regulation (PAS) 1192-2:2013 cannot be used.

[3]BS 1192-2:2013—Single source of information for any given project, used to collect, manage and disseminate all relevant approved project document for multidisciplinary teams in a managed process. Note: a CDE may use a project server, an extranet, a file-based retrieval system or other suitable toolset.

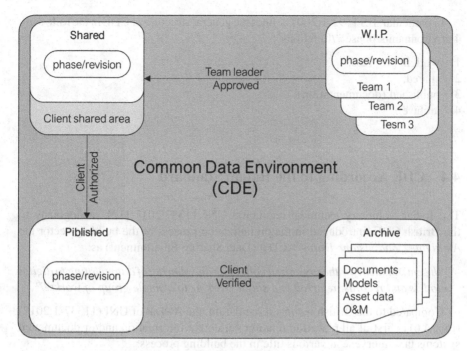

Fig. 4.4 Scheme of CDE defined in BS 1192-1:2007

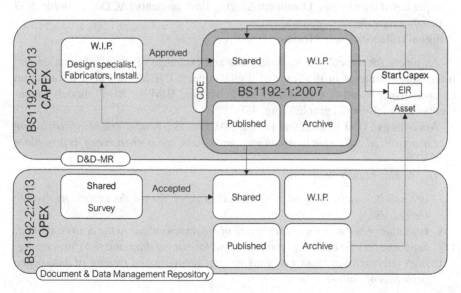

Fig. 4.5 Scheme of CDE defined in BS 1192-2:2013, digital extension and of process, execution phase

In particular, BS 1192-1:2007 defines the general structure of CDE in the following 4 environments/phases[4]/functions[5]:

1. work in progress;
2. shared;
3. published (documentation);
4. archive.

4.4   CDE According to the Italian Standard

The Italian voluntary technical regulation UNI 11337:2017 [12], analogously to the British CDE, introduced in the digitalization process of the building sector the *Ambiente Condivisione Dati*—ACDat (Data Sharing Environment) as:

– *"the environment for the organized collection and sharing of data relating to digital models and records, referred to a single work or to a single group of works."*[6]

Compared to the British technical regulation, the ACDat of UNI (11337-1:2017) (Fig. 4.6) is first of all placed in relation with other information and/or documental systems that intervene at various title in the building process:

– libraries of objects;
– paper based data rooms, Document Sharing Environments (ACDoc—*Ambienti di Condivisione dei Documenti*);
– digital collaboration platforms.

As places for the collection, processing and exchange of: data, information and information contents in the form of digital objects ("BIM" objects), information records (paper based or digital documents but not "BIM"), "BIM" models (documental and multimedia graphs) (Fig. 4.7):

According to UNI 11337:2017 [13], the ACDat *"is a computerized infrastructure for the organized collection and management of data, comprehensive of its procedure of use.*

The ACDat requirements are:

(a) *accessibility, according to pre-established rules, for all the actors involved in the process;*
(b) *traceability and historical succession of the reviews made to the data contained;*
(c) *support of a vast range of typologies and formats of data and their processing;*
(d) *high interrogation flows and easy access, recovery and mining of data (open protocols of data exchange);*

---

[4]BS 1192-1:2007.
[5]PAS 1192-2:2013.
[6]UNI 11337-1:2017.

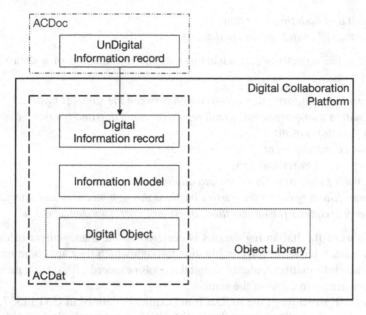

Fig. 4.6 Scheme of the building information process UNI 11337-1:2017

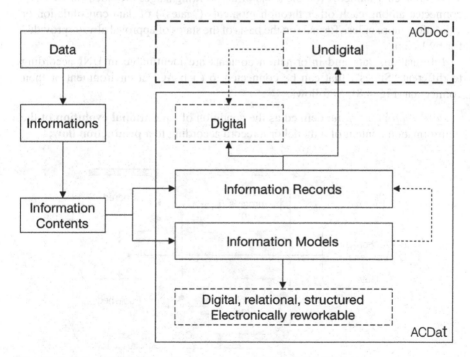

Fig. 4.7 Scheme of the building information process UNI 11337-1:2017

(e) *deposit and updating over time;*
(f) *guarantee of confidentiality and safety."*

Whereas, the objectives and advantages of the introduction of a data sharing environment are:

1. *"automation of information coordination between the subjects involved;*
2. *information transparency also with regard to ownership and temporal availability of the information;*
3. *automated management of data reviews and updates;*
4. *reduction of data redundancy;*
5. *reduction of risks associated to data duplication;*
6. *communication between the parties involved through modules and interfaces of reference (requests for information, questions, correspondence, etc.)."*

The focus of the Italian regulations is centered on the management of the single atoms (data), whatever the combination through which they are communicated (information, information contents—digital models or records, files, etc.), more than on the "container" (means) of the same.

The internal structure of the ACDat is not explicitly defined in UNI 11337:2017. On the basis of the indications of the British CDE, it can be similarly divided into 4 evolution environments (on the basis of the working stages provided for by UNI) connected among each other through moments ("gates") of data consolidation or validation, more or less formal (on the basis of the states of approval always provided for by UNIs).

Information, data and information contents are identifiable in UNI according to different "States", that can be connected to each ACDat environment or their connection (Fig. 4.8), as follows:

− *work in progress state*: defined as the condition of **operational evolution** of the information content of a model or a record according to a **production flow**;

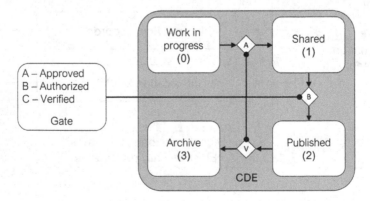

Fig. 4.8 Division of ACDat in 4 environments of information evolution, clockwise flow

– *approval state*: defined as the condition of **formal evolution** of the information content of a model or a record according to a **procedural flow** .

Therefore, depending on the operational evolution of the data production, it is possible to associate to each ACDat environment the following **work in progress state**:

L0   *in phase of processing/updating*: the information content is in the processing phase and, therefore, it can still undergo changes or updates. It is possible for the content to be unavailable for other subjects apart from the subject responsible for the production of the datum.

L1   *in phase of sharing*: the information content is considered complete for one or more disciplines, but still susceptible of interventions by other disciplines or professionals. The content is made available for other subjects apart from the Appointed party.

L2   *in phase of publication*: the information content is active, but concluded, and no subject involved apart from the Appointed party manifests the need to make any further interventions.

L3   *archived*: the information content concerns a non-active version connected to a concluded process, differentiated in:

    L3.V   "valid," version still in force;
    L3.S   "passed," with regard to versions prior to the one in force and therefore replaced.

Dissimilarly, due to a formal evolution in the data management (information or information content), in each ACDat environment and in the transit between the various environments ("gates"), the information can be associated to the following **approval state** (Fig. 4.9):

A0   *to be approved*: the information content has not undergone the approval procedure yet;

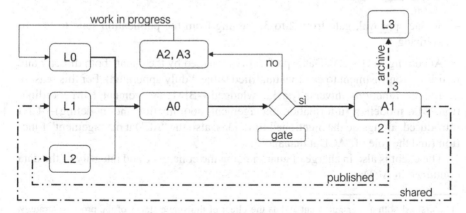

Fig. 4.9 Flow of the work in progress and approval states

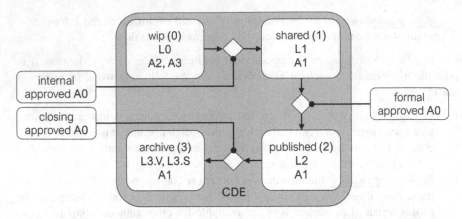

Fig. 4.10 Work in progress and approval states, according to various levels, applied to the ACDat structure

A1 *approved*: the information content has undergone the approval procedure and has obtained a positive outcome;

A2 *approved with comment*: the information content has undergone the approval procedure and has obtained a partially positive outcome, with indications relating to binding changes that the content must undergo for the following project development and/or specific uses for which it has been considered approved;

A3 *not approved*: the information content has undergone the approval procedure and has obtained a negative outcome and, therefore, has been rejected.

UNI identifies various "states" of approval but does not differentiate the possible "levels" of approval. Following the logic of the work in progress states and of the ACDat environments it is possible to assume at least 3 levels of approval (Fig. 4.10):

- internal approval, gate from 0 to 1, passing from the work in progress state to that of sharing;
- formal approval, gate from 1 to 2, passing from the sharing state to that of publication;
- closing approval, gate from 2 to 3, passing from the publication state to that of archiving.

According to UNI, ACDat is preferably managed by the client[7] both directly and indirectly (entrustment to external qualified subject duly appointed). For this reason, besides the functions universally acknowledged—BIM management, BIM coordination, BIM modeling (information management, coordination and modeling)—UNI introduced, alongside the mentioned functions, also the "ACDat management" function (and the role of "ACDat manager").

The client is also in charge of guaranteeing the coherence and integrity of the data contained in ACDat.

---

[7]Understood both as general client and as the client of the single stages of the process: strategy, planning, construction or execution (one or more stages).

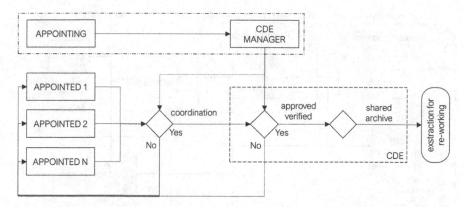

Fig. 4.11 Scheme of ACDat in UNI 11337-5:2017

The ACDat management rules are specifically defined in the Information Tender Specifications (CI—*Capitolato Informativo; UK/ISO* Employer/Exchange Information Requirements—EIR),[8] including the ACDat ownership (or custody) at the end of each stage or phase of the process, should the originating client be changed. According to UNI, only in particular cases, ACDat can be delegated in the CI (as construction and management) to one of the appointed parties.

According to the Italian regulations, therefore, given its function as place that guarantees data sharing and integrity, and not only as mere documental container, the creation and management of ACDat, also when indirect, is ascribable to the Appointed party only in particular cases, remaining preferable to leave it under the client's responsibility.

The CDE scheme provided in the regulation (11337-5:2017, Fig. 4.7, par. 7) represents the idea of a basically single environment for the entire process, from the strategic planning of a work up to its execution. The aim is to optimize data processing, coherence and interaction through the uniqueness of the same and limited data redundancy (Fig. 4.11).

By simplifying the British CDE scheme, UNI's scheme focuses the attention on the information flow.

Considering that BSI 1192-1:2007 is the regulation on British design, it is evident that the relevant CDE assumed is a CDE of the design Team where the client interacts in the sharing and publishing spaces (Shared-Client shared area and Published-Client shared area).

---

[8]UNI 11337-5:2017:

– Informative Tender Specifications (CI—Capitolato Informativo): Clarification of needs and information requisites required by the client to the Appointed party.

Note—The Information Tender Specifications correspond, in their essential lines, to the Employer Information Requirement (EIR) of PAS 1192-2:2013.

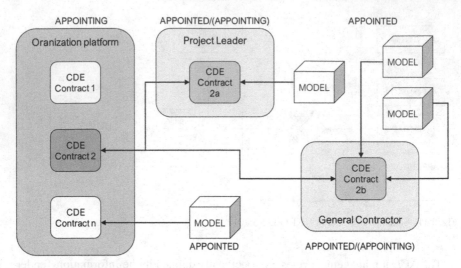

Fig. 4.12 CDE flow

This, together with the concept of uniqueness of the sharing environment, led to think in the beginning—and still today for some that are not very expert or informed—that the CDE could be a single environment left in charge of the designer or, subsequently, of the building company.

Therefore, it is the case to clarify several aspects.

The CDE is a data sharing environment that an organization opens for each specific job order. The various job orders (and related CDEs) are then organized centrally in the organization's collaboration platform. Therefore, the CDE job order is always under the client's responsibility as owner of the results (works, services, supplies and, of course data) of the job order (Appointing Client, Owner) [6].

The various subjects that will intervene in the job order (Appointed party) will be able to interact directly in the CDE job order or in the event of a general contractor in the CDE design or construction (or of both, Lead Appointed party) through the respective CDE contracts (where the latter are clients of their suppliers or subcontractors) (Fig. 4.12).

The Appointing and Lead Appointed parties operate in their own WIP[9] and make their models and documents available to the other parties in the Sharing environment (for coordination) or in the Publishing environment (for deposit or sharing of the completed work). The appointed party approves the documents processed in WIP and makes them visible in the sharing area for the coordination with all the parties involved (in case, the appointed party can also invite the appointing party for the latter's initial considerations). Once the coordination is concluded, the appointed

---

[9]Only in this way the various WIPs remain reserved to the sole compilers of the information. Although filtered, the access to each part of the CDE (as for all DBs) is always open to the System Manager, which allows to have a truly reserved WIP, with the exclusion of who is also System Manager.

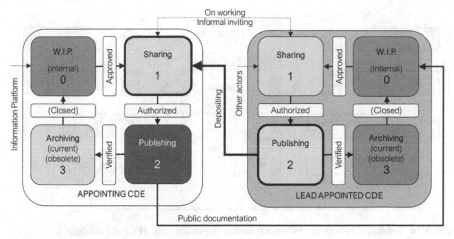

Fig. 4.13 Appointing CDE (Client) and Appointed CDE (Tender)

party authorizes the publication (models and documents can be reused by others), and for the appointed party they are ready to be deposited in the Appointing party's CDE.

Therefore, the CDE managed by the Appointing Party receives the definitive data coordinated in advance between the Appointed parties in the Lead Appointed Party's CDE. The entry of the Appointed parties' coordinated data in the CDE job order takes place in the Sharing environment, it assumes value of deposit for the client, and is always subject to the client's approval and verification. The CDE data—coordinated, approved and verified, upon the client's prior authorization—can therefore be shared with third parties (Publishing) or in conclusion archived (Archive) for their following possible mining, should they be re-processed (Fig. 4.13).

4.5 Structure of CDE Data

The information content of ACDat is represented by data and by data structures (files or generally speaking containers—PAS 1192-2). It represents something different from a file archive and sharing system as well as from a documental manager, despite carrying out the same activities. In addition to which, there are those relating to the management of the workflow (BPM), the visualization of general files and graphic models (in open and native format), the management of the planning, etc (Fig. 4.14).

Generally speaking, it concerns relational Databases, but there are ACDat prototypes based on non-relational DBs [8]. Just as in commerce there are cloud systems (SaaS or PaaS) and stand-alone systems (on premises) without distinction.

Its main objective is to guarantee over time, and according to specific rules, the digital sharing of the data generated by the various subjects of a building process.

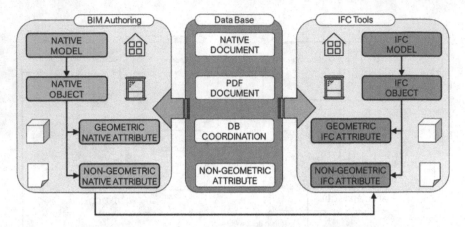

Fig. 4.14 Relationship between ownership formats, open formats (IFC) and database

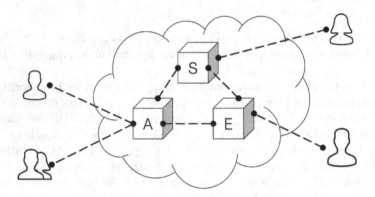

Fig. 4.15 Collaboration platforms for the drawing up of models

Its origin for design purposes focused its use within design teams with the aim to share graphic model files (generated by BIM authoring software) drawn up by the various disciplines: architectural, structural, mechanical plant-engineering, hydraulic, electric models, etc. (Fig. 4.15).

Issues relating to the coordination of (graphic) files, federation of models, verification of geometrical interferences (clash detection), processing of files at the same time by more operators, etc. (see BIM Plus, Trimble Connect, Collaboration for Revit, BIM 360 Team, BIMX, etc.).

Today, for these aspects, there are specific platforms connected, generally speaking, to the same software houses that produce BIM authoring programs which, over time, have generated or are evolving into actual ACDat/CDE (as extension of the former or as distinct platforms). These are ACDat/CDE in which the original functionality extends, with regard to collaboration on graphic models, to the management of data, files and documents of different nature (relations, calculations, computations, technical datasheet, etc.), and to the coordination not only of 3D models, but also of aspects relating to the management of timing, 4D, costs, 5D, maintenance, 6D, etc. [4].

4.6 Structure of CDE Environments

Particular attention must be paid also to the internal structure of ACDat single environments (Fig. 4.16).

Generally speaking, the environments are divided internally on the basis of disciplines:

- surveys;

  - site
  - manufactured articles
  - systems

- territory;

  - town planning
  - urbanizations
  - green
  - infrastructure

- architecture;

  - construction
  - finishing

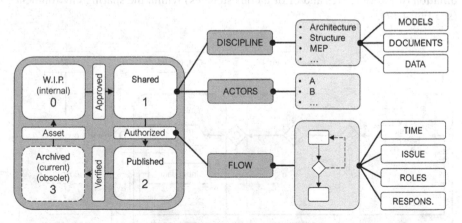

Fig. 4.16 View of the elements composing the ACDat as disciplines, subjects and flows

- façades;
- structures;
- systems;
  - mechanical
  - electrical
  - hydraulic
  - air conditioning
  - transport

- safety;
- fire-prevention;
- hygiene;
- acoustic;
- energy efficiency;
- cadastre;
- etc.

The flexibility of the computerized system chosen allows to define the structure of the environments on the basis of each work, or on the basis of the process phase and of the relevant specific needs of data organization.

4.7   Definition of Workflows

The same attention must be addressed to the definition of the process flows deriving, for example, from the transit of models and records throughout the various "gates" of approval between environments (internal, formal, of closure) or from their coordination (if interferences and/or of inconsistencies) within the sharing environment (see Fig. 4.17).

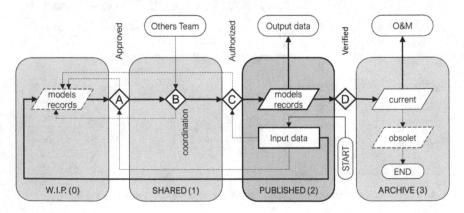

Fig. 4.17 CDE base workflow

In particular, for each workflow it is necessary to define the figures involved, the persons in charge, the activities planned and the consequent ones, the timeframe, the possible correction actions, etc.

- Gate A internal approval (Fig. 4.18):

  - Transit from the Work in progress environment (0) to the Sharing environment (1);
  - Models or records from the Work in progress state (L0) to the Sharing state (L1);
  - Internal approval: to be approved (A0), Approved (A1), approved with comment (A2), not approved (A3);
  - Verification: formal internal (V1), substantial (V2);
  - Coordination: within a model (LC1).

- Gate B coordination (Fig. 4.19):

  - Progress made within the Sharing environment (1);
  - Coordination: among models (LC2).

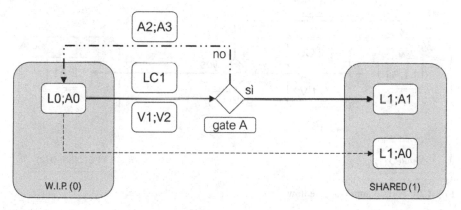

Fig. 4.18 Flow of the internal approval gate

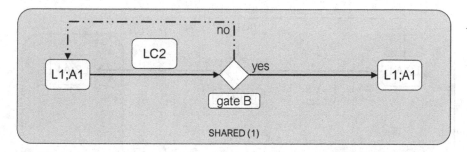

Fig. 4.19 Coordination gate flow

- Gate C formal approval (Fig. 4.20):

  – Transit from the Sharing environment (1) to the Publication environment (2);
  – Models or records from the Sharing state (L1) to the Publication state (L2);
  – Public approval: to be approved (A0), Approved (A1), approved with comment (A2), not approved (A3);
  – Verification: formal internal (V1), substantial (V2);
  – Coordination: models and records (LC3).

- Gate D closing approval (Fig. 4.21):

  – Transit from the Publication environment (2) to the Archive environment (3);
  – Models or records from the Publication state (L2) to the Archiving state (L3);
  – Public approval: to be approved (A0), Approved (A1), approved with comment (A2), not approved (A3);
  – Verification: formal and substantial external (V3) if provided for, anyway formal internal (V1) and substantial (V2).

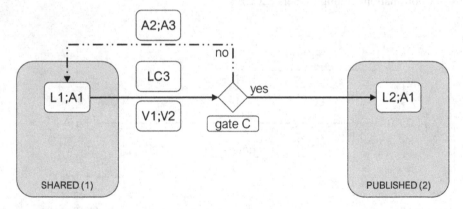

Fig. 4.20 Formal approval gate flow

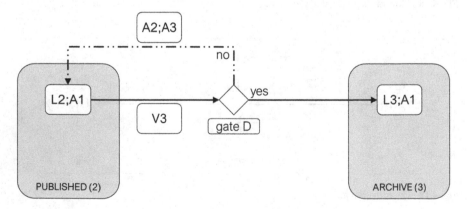

Fig. 4.21 Closing approval gate flow

4.8 CDE in the Management of Contracts

A specific need of CDE is to host and manage the material produced by third parties both as deposit environment and as verification and acceptance environment of the material stored, increased with the additional data entered by the CDE deposit manager. This function of course is significant for all clients (contracting stations, etc.) and can assume specific restrictions in case of public contracts.

A first aspect, in general but especially in the period of complete transition to models, concerns the contemporary presence and congruence of the graphic model data (the only one in this moment that can truly be required: level 2 digital maturity, elementary) and every other datum contained in the remaining records (extrapolated or not extrapolated from the models)

Moreover, the CDE has different uses and flows depending on whether it involves the "tender" phase or the following phase of the actual execution of the service or contracted work.

4.8.1 CDE in the Tendering Phase

The first use of a management and deposit CDE of tender models and records involves the tendering and awarding phase. In this case, the deposit concerns: the tender in general, including any annexed documents, and, especially, the "**pre contract BIM Execution Plan**" including any annexed records and models (Fig. 4.22).

In particular, for each workflow, as always, it is necessary to define the figures involved, the persons in charge, the activities planned and the consequent ones, the time, the possible correction actions, etc.

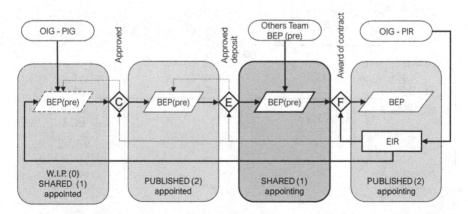

Fig. 4.22 CDE flow in the tender and awarding phase

- Gate C formal approval:
  - Transit from the Sharing environment (1), Appointed, to the Publication environment (2) Client;
  - pre contract BIM Execution Plan (BEP(pre)) from the Sharing state (L1) to the Publication state (L2);
  - Formal approval of the appointed party: to be approved (A0), Approved (A1), approved with comment (A2), not approved (A3);
  - Verification: formal internal (V1), substantial (V2);
  - Coordination: between BEP(pre) and annexed records or models (LC3).

  - Gate E approval for the deposit (Fig. 4.23):
    - Transit from the Publication environment (2), Appointed, to the Sharing environment (1), Client;
    - pre contract BIM Execution Plan (BEP(pre)) from the Publication state (L2) to the Sharing state (L1), with regard to other tenders;
    - Similarly to the above gate C, approval of the person in charge of the job order.

- Gate F awarding (service, work, supply) (Fig. 4.24):

  - Transit from the Sharing environment (1), Client, to the Publication environment (2) Client;
  - pre contract BIM Execution Plan (BEP(pre)) of the Sharing state (L1), post contract BIM Execution Plan (BEP(post)), Publication state (L2);
  - Awarding of the contract: to be approved (A0), Approved (A1), not approved (A3);
  - Verification: formal internal (V1), substantial (V2);
  - Coordination: between BEP(pre) and annexed records or models (LC3).

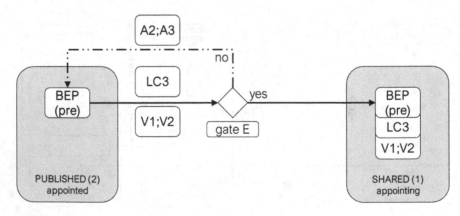

Fig. 4.23 Depositing of BEP (pre) flow

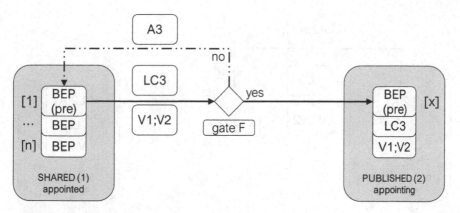

Fig. 4.24 Award of contract flow

N.B.: the awarded "*pre contract BIM Execution Plan*" in the Sharing state is brought to the Publication state to be confirmed through the "*post contract BIM Execution Plan*" (post-BEP).

4.8.2 CDE in the Contract Phase

The second use of a management and deposit CDE of contract models and records concerns the execution phase of the contract (services, works or supplies). In this case, the object of deposit involves models and records produced by the Appointed party (Fig. 4.25).

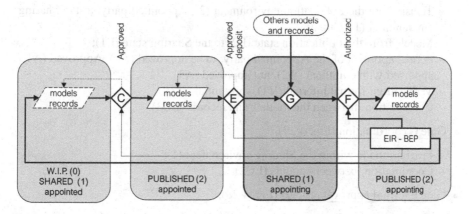

Fig. 4.25 Tender execution CDE flow

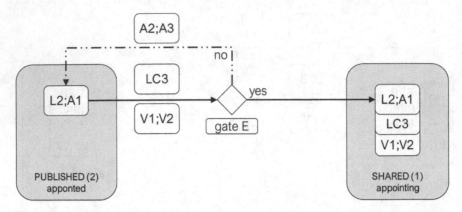

Fig. 4.26 Approval gate flow

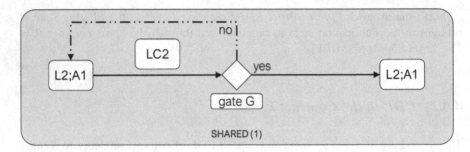

Fig. 4.27 Coordination gate flow

- Gate E approval for deposit (Fig. 4.26):

  - Transit from the Publication environment (2), Appointed party, to the Sharing environment (1), Client;
  - Models from the Publication state (L2) to the Sharing state (L1);
  - Formal approval of the Appointed party: to be approved (A0), Approved (A1), approved with comment (A2), not approved (A3);
  - Verification: formal internal (V1), substantial (V2);
  - Coordination: between the models and records (LC3).

- Gate G coordination (Fig. 4.27):

  - Progress made within the Sharing environment (1);
  - Coordination: between models (LC2).

- Gate F formal approval (client) (Fig. 4.28):

  - Transit from the Sharing environment (1), Client, to the Publication environment (2), Client;
  - Models or records from the Sharing state (L1) to the Publication state (L2);

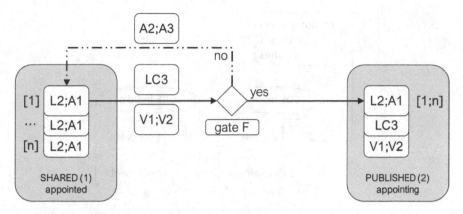

Fig. 4.28 Formal approval gate flow

- The client's approval: to be approved (A0), Approved (A1), approved with comment (A2), not approved (A3);
- Verification: formal internal (V1), substantial (V2);
- Coordination: models and records (LC3).

4.9 Digital Platform

In the era of data management (of Big Data and Smart Cities), a concluded tool for sharing processes within a single job order or in relation to a single asset (building or infrastructure), such as the CDE, results to be also a tool that limits the potentialities introduced by BIM methodologies and tools if not integrated in a system of further higher coordination.

For this reason, today's tendency is to increasingly refer to data management "platforms" (UNI 11337:2017), that overcome the limits of the current CDE (relating to the single building or the single intervention), as places for the collection, management and installation of all the information coming from objects, models and various CDEs (Fig. 4.29).

Big and small properties (public or private), managers, contracting stations, etc., multi-tenant, can thus install not only the single asset or the single intervention, but also the integrated management of their entire real estate portfolio. Likewise, in parallel, credit, insurances, investors, production, etc. can draw precise and scenario information from the sector platforms and the installation of the data concerning the entire sector and territory, in order to make relevant strategies, optimizations, reallocations, etc.

As of 2013, different sector digital platforms have been developed, BIM oriented, also at national level. With different profiles in their midst (BIM library, BIM server,

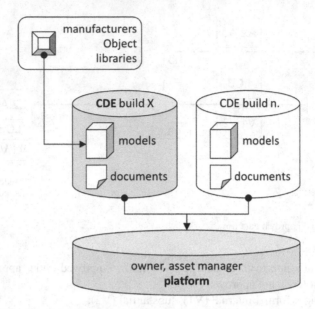

Fig. 4.29 Information management flow between: libraries of objects, models, CDEs and platforms

Table 4.1 National and European digital platforms

Date	Name	Nationality	Description
2013	INNOVance	Italy	BIM Library for manufacturer object, standard attribute; BIM Server for models and project, tenant management; BIM-GIS platform for building, infrastructure and site
2015	BIMToolkit NBL Library	United Kingdom	BIM Library; Standard definition of geometrical an information attribute (LOD-LOI); valuation Toolkit of object LOD and LOI level in a model
2017	Kroqi	France	BIM Server, design oriented; software and API collector for project collaboration with commercial tools solution

etc.), but univocal in the objectives: to manage and coordinate the data of the building sector besides the single interventions or single assets [2] (Table 4.1).

Basically, the national platforms have the objective to foster the establishment, development and dissemination of digitalization in the building sector. On the one hand, their aim is to centralize the rules and part of structure costs and investments, difficult to bear especially for small realities with scarce personnel and not much capitalized. On the other hand, they stimulate the use of open formats, not proprietary, and competition among software houses, in order to limit dominant positions, or monopolies, as those currently present, for example in the CAD environment.

4.10 INNOVance Platform

Compared to the other platforms that mainly deal with a specific aspect, INNOVance [7] covers the building process on a full scale. As analyzed in detail in the previous chapter, INNOVance is a collaboration platform that collects in a structured way and puts in relation all types of information at the user's service and at the service of the latter's software that, through open protocols, can operate on it extracting and re-depositing rough and/or aggregated data, improving its process owing to the continuous increase of the knowledge contained, of whatever type it may be.

Therefore, INNOVance is the ideal conceptual basis for the future European platform because it is not limited to "BIM" models and objects, but it manages data and information on:

- products (from sand, to mortar or brick, central heating such as boiler, glass and window);
- subsystems (from the layer of plaster to that of masonry or of insulation, water resistant, etc.);
- systems (packages of closing or division wall, packages of division or covering floor, etc.);
- works (houses, industrial structures, green, infrastructure, etc.);
- spaces (from residential or tertiary volumes, to hot/cold/fireproof areas, environments, rooms, etc.);
- works (substructure, construction, installation, maintenance, etc.);
- human resources (professionals, employees, specialists, etc.);
- means and equipment (cranes, excavators, concrete mixers, such as grinders or drills, etc.).

Therefore, in INNOVance it is possible to research, manage and download the BIM object, the video showing the installation of a product, as well as tolerances, control activities or works preparing support, the nature and typology of a typical team, the type of patent necessary for the operator with regard to a specific machine, as well as the operator's safety devices, the type of packaging and the quantity of waste generated and the necessary disposal operations, the timeframe of installation as well as the periods of programmed maintenance, the location of the production plant, incompatible products, legal dimensions of a room in accordance with the regulations of reference, the costs of a product—such as of the execution of a work up to the accounts of a work—its volume, the gross floor area, urbanization expenses, analyses of the ground, cadastre, the load and movement of a crane, the power of an excavator and the type of accessories that can be installed: bucket, compactor, pliers, etc. and all their characteristics.

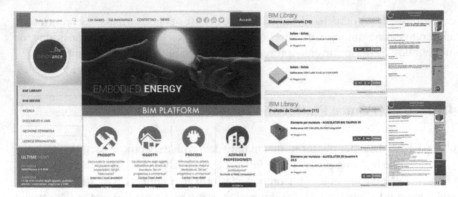

Fig. 4.30 Front-end of the Innovance portal with the various options of use

In INNOVance not only it is possible to "download" an object and enter it in its model with all its attributes, as in usual libraries, but it is also possible to build, for example, a wall defining its layers and choosing the relevant materials though product datasheets present directly on the portal, and then only at the end send everything to the BIM authoring software (Revit, Allplan, Archicad, Aecosim, etc.) so that the object wall may be modeled[10]; it is also possible to do the opposite and model a wall in the BIM authoring software and then send those the geometries and attributes to INNOVance so that it may be the platform to identify the information datasheets to which to refer in order to complete the information attributes necessary for the use of that object for the entire chain. Completion attributes that remain indissolubly connected to the objects, through the mentioned information datasheets, without burdening the graphic file (Fig. 4.30).

4.11 BIMReL Project

Different from INNOVance, but similar in the concept of centralizing data and not geometries, BIM Rel is the answer provided by Politecnico di Milano to the common Libraries of objects, public or commercial.

BIM ReL is the regional library of BIM digital objects aimed at creating a digital display on the global market for enterprises in Lombardy, above all PMI, co-financed through the tender Smart Living of the Region of Lombardy. Politecnico di Milano is the leader of the project (2017–2019), together with industrial partners One Team and Trace Parts.

The aim of BIM ReL is to favor producers of component products so that constructions and systems plants can enter the international market through digital innovation.

---

[10]The functionalities of self-composition of the objects starting from the information coming from INNOVance were tested with Autodesk's Revit 2013.

BIM ReL places the product, its characteristics and performances in the centre, using 3D geometry as a reading means, a driver of work information, more than as the element of "attention" and "decoration" for the sole purpose to complete the graphic model. This, instead, it was often occurs with the other "BIM" libraries, that are full of objects of fittings and finish more than of actual construction components; they are also full of more or less complex and rendered 3D geometries instead of alphanumeric attributes necessary in the construction and execution phases.

BIM ReL is a library of BIM objects that starts from the product's technical regulations and from the essential requirements of CPR 305/2011, and the EC marking, to describe an element that enters the "downright" process of constructions from its actual digital virtualization, even before as physical component in, and of, the final building/infrastructure product (Fig. 4.31).

Therefore, the EC marking, and anyway the technical regulations of the sector, become the basis for the transparency and readability of requirements, the marketability and quality of products, a guarantee for the consumer/user. BIM ReL follows the same path chosen both within the scope of UNI (UNI 11337 part 3), and within the scope of CEN (TC442, WG4) [3], to define the attributes that identify and describe the construction and system digital objects, for the future European BIM product datasheets and digital marking, thus departing from the usual concept of geometry libraries (Fig. 4.32).

The information attributes identified in the standard of reference (UNI 11337-3:2015) were in part confirmed also in the regulatory works of CEN table 442 as synthesized in Table 4.2.

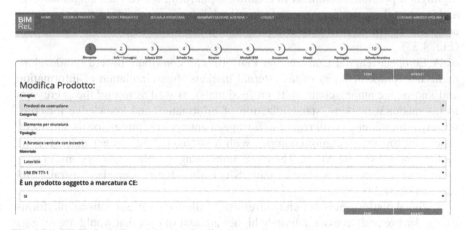

Fig. 4.31 BIMReL platform (object uploading process)

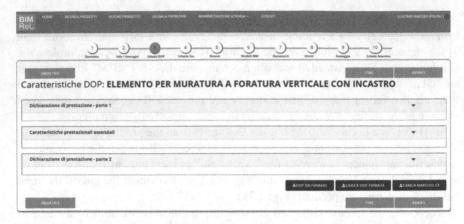

Fig. 4.32 BIMReL alphanumeric attributes

4.12 European Digital Platform

It is currently evident that the market needs portfolio platforms at a higher level of the single CDEs (relating to single assets or single job orders) and national platforms above both. Likewise, it is also evident that a supranational—community—common market very likely presents the same critical aspects related to data collection, sharing, transparency and management of the lower levels. These critical aspects highlight a possible usefulness in creating a platform, above each single national and local market, with a continental nature and dimension capable of fostering interactions between professionals and information flows also beyond state boundaries (Fig. 4.33).

A European Digital Construction Platform can contribute toward the development and competitiveness of the internal markets (free circulation of information and knowledge among community professionals), as well as toward the aggression to international markets (standardization of information and knowledge for community professionals) with regard to European enterprises, professionals and users, especially toward emerging countries with a stronger development index for the sector, that is Asia and Africa. Therefore, competing as a European system (in constructions) against giants such as the USA, China, Japan, Canada, Australia and Russia (Fig. 4.34).

A digital platform at community level would allow the various national platforms to standardize and access a definitely higher amount of data that would improve its current performances increasing the services offered without additional costs. At the same time, it would allow functions of massive data mining indispensible to define a common language in the sector at continental level (also overcoming the current language barriers) and the definition of standard information attribute structures for buildings as well as for construction products (digital marking of EC products; Product Regulations 350/11; Afnor PR XP P07-150 [1]; UNI 11337-3:2016 [11]).

Table 4.2 UNI and CEN (Smart CE) object attribute

UNI 11337-3:2016	CEN 442 (SmartCE)
General Information	General information block
Identifying manufacturer information	
Identifying product information	
Technical information	
Morphological and descriptive features	Declared performance
Geometry and shape	Dimensions and dimensional tolerances
Appearance and constructive features	Configuration
Dimensions	Compressive strength
Physical and chemical properties	Cond strength
Qualitative	Reaction to fire
Quantitative	Water absorption
Tolerance	Water vapour permeability
Main components of the product	Direct airborne sound insulation
Declared performance characteristics	Thermal resistance
Essential characteristics	Durability against freeze/thaw
Voluntary characteristics	Release of dangerous substances
Information about sustainability according to UNI EN 15804	Appropriate and specific technical documentation
Safety information	Signature
Information about packaging, movement, deposit in factory and transport	Additional information related to the DoP
Commercial information	Compressive strength
Additional technical information	Bond strength
Supplementary documents	
Attachments	
Information on data reliability	

The platform would allow the production of knowledge useful for the development of new technical voluntary standards for the CEN—and through the latter for ISO—as a community contribution, and not only national, of the single member states (Reg. 1025/12: *"The European regulations also contribute toward the promotion of competitiveness among enterprises fostering in particular the free circulation of goods and services, network interoperability, means of communication, technological development …"*) [5].

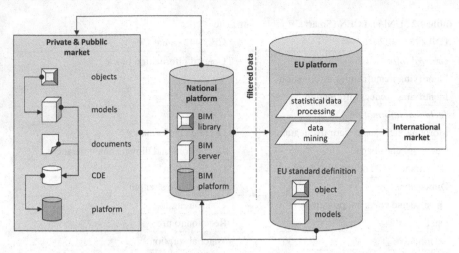

Fig. 4.33 Objects, models, documents, CDE, platforms and Data flow in private and public markets to national and European digital platforms

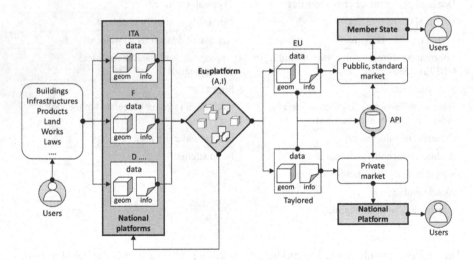

Fig. 4.34 European digital construction platform

4.13 Conclusions

The digitalization of the building industry allows the sector to fully and rightfully enter Industry 4.0 as any other industrial sector and of services.

A characterizing aspect of this fourth revolution is data management and enhancement. In the building sector, the digital management of information involves objects and BIM models, the Common Data Environment, digital platforms of the sector (real estate portfolios), and national platforms.

The development of a supranational platform would allow the building sector to overcome the current barriers of language, knowledge, national and local regulations, allowing also to define structures of common information attributes standardized for construction products (digital EC marking; CPR 305/11).

A continental digital platform would favor community professionals in the internal market and the aggression of the international ones (above all, Asia and Africa) as a European system of the building industry with regard to giants such as: the USA, China, Japan, Canada, Australia and Russia.

A community market platform would provide the Member States with an open and controlled infrastructure, source of indispensible data also for the public market, widening and strengthening the functionalities of the single national platforms and of commercial software solutions. It would give access to data coming from all over Europe without bearing launching costs and especially maintenance and updating costs. Moreover, the European platform would spur the Member States still not provided with national platforms to equip themselves in such sense, accelerating the digitalization process of the building sector.

References

1. Afnor (2014) AFNOR XP P07-150, Properties of products and systems used in construction—definition of properties, methods of creating and managing properties in a harmonized system of reference. AFNOR
2. BIMobject Corporate (2019) BIM object. https://www.bimobject.com/it
3. CEN/CT442/WG04 (2018) CEN CWA 17316:2018—Smart CE marking for construction products. CEN, Brussels, pp 61. https://www.cen.eu/news/brief-news/Pages/NEWS-2018-025.aspx
4. Department of Administration - Division of Facilities Development (DFD) (2012) 'Building Information Modeling (BIM) Guidelines and Standards for Architects and Engineers'. https://doi.org/10.1097/acm.0b013e3181e8dbca
5. European Commission (2011) Regulation (EU) No 305/2011 of the european Parliament and of the Council, Official Journal of the European Union, Strasbourg. https://eur-lex.europa.eu/LexUriServ/LexUriServ.do?uri=OJ:L:2011:088:0005:0043:EN:PDF%0Apapers3:/ /publication/uuid/BD5A1031-4234-4755-80B7-C824AAEA8F82
6. ISO/TC59/SC13/WG13 (2019) BSI EN ISO 19650-1:2019. Organization and digitization of information about buildings and civil engineering works, including building information modelling (BIM)—information management using building information modelling—PART 1: concepts and principles. BSI Standards Limited, London, pp 1–46
7. Pavan A et al (2014) 'INNOVance: Italian BIM database for construction process management. In: Computing in civil and building engineering—proceedings of the 2014 international conference on computing in civil and building engineering. https://doi.org/10.1061/9780784413616.080
8. Solihin W et al (2017) A simplified relational database schema for transformation of BIM data into a query-efficient and spatially enabled database. Autom Const 84:367–383. https://doi.org/10.1016/j.autcon.2017.10.002 Elsevier
9. The British Standards Institution (2013) PAS 1192-2:2013, Specification for information management for the capital/delivery phase of construction projects using building information modelling. British Standard Institute. British Standard Limited, UK. ISSN 9780580781360/BIM Task Group

10. The British Standards Institution (2014) PAS 1192-3:2014, Specification for information man-
    agement for the operational phase of assets using building information modelling. British
    Standard Institute. British Standard Limited, UK. ISSN 9780580781360/BIM Task Group
11. UNI, CT033, WG05 (2015) UNI TS 11337-3:2015, Building and civil engineering works—
    codification criteria for construction products and works, activities and resources—Part 3: mod-
    els of collecting, organizing and recording the technical information for construction products.
    UNI, Milano, p 32
12. UNI/CT033/WG05 (2017) UNI11337-1:2017, Building and civil engineering works—digital
    management of the informative process—Part 1: models, documents and informative object
    for products and processes. UNI, Milano, p 26
13. UNI/CT033/WG05 (2017) UNI11337-5:2017, Building and civil engineering works—digital
    management of the informative process—Part 5: informative flows in the digital processes.
    UNI, Milano, p 24

Chapter 5
Benefits and Challenges
in Implementing BIM in Design

Abstract The acronym BIM has been linked to several interpretations in the different phases of its maturity considering the tools, the processes, the method, etc. To comprehend the challenges in BIM implementation it is crucial to understand the context and the limitations that nowadays are shaping the boundaries of its development. On the one hand, this chapter describes some of the technological challenges in BIM application starting from the concept of object-oriented programming. It presents, among the others, a critical discussion about the time and cost integration in BIM (4D and 5D) and the use of open languages (such as IFC). On the other hand, the chapter explores the context of BIM application studying national and international standards and legislations, providing a map of the existing standards and their relations. Finally, a discussion about the evolution introduces by the ISO 19650 is proposed including a novel interpretation of the passage from LOD to LOIN.

5.1   Introduction

"BIM" is defined in many ways:

ISO 29481-1:2010
"2.2 building construction information model—BIM
shared digital representation of physical and functional characteristics of any built object (including buildings, bridges, roads, etc.) which forms a reliable basis decisions
Note: "Building Information Model" is frequently used as a synonym for BIM"

PAS 1192-2:2013
"3.7 building information modelling (BIM)
process of designing, constructing or operating a building or infrastructure asset using electronic object oriented information"

© Springer Nature Switzerland AG 2020
B. Daniotti et al., *BIM-Based Collaborative Building Process Management*, Springer Tracts in Civil Engineering,
https://doi.org/10.1007/978-3-030-32889-4_5

ISO 29481-1:2016
"3.2 building information modelling—BIM
use of a shared digital representation of a built object (including buildings, bridges, roads, etc.) to facilitate design, construction and operation processes to form a reliable basis for decisions
Note 1 to entry: The acronym BIM also stands for the shared digital representation of the physical and functional characteristics of any construction works"

EN ISO 19650-1:2018
"3.3.14 building information modelling—BIM
use of shared digital representation of a built *asset* to facilitate design, construction and operation processes to form a reliable basis for decisions
Note 1 to entry: Built assets include, but are not limited to, buildings, bridges, roads, process plants."

Over the years, the meaning of the acronym (BIM) has changed too: Building Information Model, Modeling, Management. Even the actual acronym has been modified from "BIM" to "CIM" (Construction), "HBIM" (Heritage), "EIM" (Environmental), etc.

To understand a new science, knowledge, method or tool, we must understand the real meanings of its terms [16].

5.2 BIM Method or Tool

Among the academics, the debate has especially concentrated on defining BIM's nature, that is whether it is as *method* or a *tool*. Whereas, with regard to its practice, the most recurrent terms to describe BIM still are: 3D model, parametric model (3D), BIM model, object modeling and a vague reference to the concept of "database" (not better specified), or a more or less complex variation of all the above put together.

It is possible to state that the "doctrinal" prevalence over the abovementioned is to view BIM as a "method," even if often understood/confused in substance as a mere collaboration approach between subjects and disciplines, and as a more "operational" approach, thus as a "tool," that tends to confuse it among the computerized tools for developing and managing graphics and geometries (the so-called BIM "authoring" software), up to its identification in only one of these (the most common at commercial level).

Analyzing the information available, it is also clear that:

– the collaboration nature of subjects and tools involved in the building process is a need of the sector prior to BIM (that can certainly be favored by this aspect);
– the 3D representation software has long-existed without being BIM (CAD 3D, 3D modelers, rendering, etc.);
– the use or the possible connection to a Database is by now common to all CADs and not certainly a prerogative of BIM;

Fig. 5.1 Object oriented programming: objects and classes

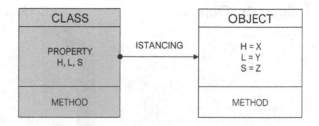

– an IFC model is parametric [3], and if this (IFC) is the objective language of the open BIM then parametricity cannot be BIM's true essence [15].

So what is BIM in actual fact?

First of all, it is the modification of the tool, from a representation system to a simulation system. Such transit took place first of all in graphics and in the management of geometries, and it can be more easily explained with graphics, also defining the limits. However, it is then to be viewed as an approach for every tool supporting process, evaluation, calculation, etc.

The first step is represented by the computerized introduction of the Object Oriented Programming (developed at academic level in the '60s—Simula—and then at commercial level in the 1980s/1990s—Java, C++, C#, Python, etc.) [2].

Simplifying the concept, it is possible to say that the object oriented programming introduces the machine to the real world, made of physical or abstract entities, that are simulated in a computerized sense. These entities can be described on the basis of property, "attributes," actions and "methods," homogeneous among each other with the possibility of being grouped into "Classes." Each time the attributes of a class are enhanced (istanced), a virtual "Object" is developed (object oriented programming), simulating reality (Fig. 5.1).

In the more recent CADs [14], whose purpose is the computerized assistance in design in terms of representation (Computer Aided Design) [19], the classes are general representation tools: lines and surfaces (2D), volumes (3D). In modern CADs, therefore, we have "taught" the machine that the world is made of lines, rectangles and parallelepipeds, and we ask the machine to operate with them [6].

Therefore, the ontology of object oriented CADs is essentially the formal representation of "geometry."

The tool: pencils, drafting machines and computers have improved and made efficient the designer's operativeness (move, cancel, copy, etc.), but in an absolutely dominated and unconscious way with regard to what those shapes represented (a wall, as well as a bolt, a tree or an airplane).

With BIM, the previous information is reset and the machine is told that the environment, or at least the AEC environment (Architecture, Engineering and Construction), is not made of geometries but of windows, walls, roofs, doors, boilers, pillars, etc. Therefore, BIM allows to simulate the building objects representing them graphically with geometries.

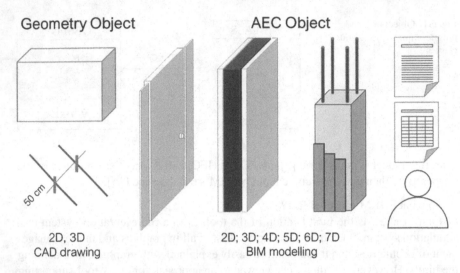

Geometry Object   AEC Object

2D, 3D   2D; 3D; 4D; 5D; 6D; 7D
CAD drawing   BIM modelling

Fig. 5.2 CAD drawing versus BIM modeling

BIM's ontology (object oriented) is thus essentially the formal representation of the building industry (AEC) [18].

Building objects that can also be entities impossible to represent graphically, but definable on the basis of attributes and methods: additives for concrete, contracts, operators, activities, etc (Fig. 5.2).

Therefore, BIM is also collaboration, 3D, parametrization, use of database, etc. but especially computer modeling. The simulation of the building environment (AEC) through a machine capable (also) of autonomously reprocessing information (object oriented programming, with building ontology/dominium).

Hence, the "method" changes although not in the sense of collaboration with people and disciplines, but in the sense of collaboration with the machine, because the "tool" has truly changed (not only evolved). For the first time in history, the tool is not only executor of actions and holder of data, but it becomes an actual collaborator of the process; it has the operators' same "dictionary" (Fig. 5.3).

Therefore, it becomes possible, through shared standards, to instruct the machine with specific building rules, to carry out verifications, etc. To perform, or better, to make perform the so-called BIM review operations: clash detection (control of interferences between geometries and room for maneuver) and code checking (control of information inconsistencies on the basis of pre-established rules—regulations, laws, contractual requirements, etc.).

Interaction with the machine obviously has limits that operators generally tend to solve with a typical design approach (and CAD) but that, if acceptable in the traditional representation systems, risks to create misunderstandings and errors in BIM modeling too often underestimated.

Just think of the class "walls," that graphically have a parallelepiped conformation, with specific masonry attributes and methods. Some BIM authoring software do not

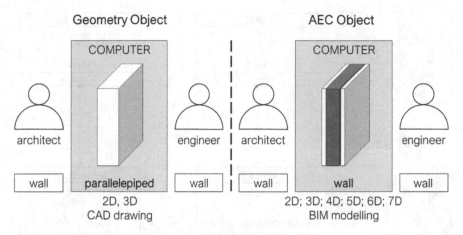

Geometry Object   AEC Object

Fig. 5.3 Computer interaction with operator

have finish complements among their classes as in the case, for example, of entities such as skirting boards, these too graphically representable with a parallelepiped but with specific skirting board attributes and methods only in minimum part similar to a masonry. In 3D CAD, the designer will use a parallelepiped for the wall and another parallelepiped, which the designer will call skirting board, to solve the problem of representing the finish element. Not finding the class "skirting board" in BIM among those in the software, the designer could be driven, for graphic similarity, to use the class masonry and to generate (istance) a masonry with particular dimensions, calling it, as previously in CAD, "skirting board." At graphical level, the designer will have solved the representation problem, but not certainly the modeling one (simulation of realities). From that moment on, what the designer sees (skirting board) will no longer be what the machine controls, a very tiny wall on top of another wall (the latter with "normal" dimensions). Also the attributes and methods will be the same of the walls (it hooks on to floors, it hosts doors and windows, etc.) and not certainly of a skirting board, which the machine, on the other hand, does not know (Fig. 5.4). What the operator sees is no longer what the machine understands, losing the peculiarity of BIM, and reasoning in a traditional way as if it were a CAD.

The same thing happens when the so-called "generic objects" are used, which regardless of the name (denomination attribute) they are given, will fall within the class of "generic" entities for the machine (with absolutely general attributes and methods) and not as they are viewed by man: spires, gables, or capitals, etc. (Fig. 5.4).

It is important not to confuse the possibility, typical of Databases, to associate/relate a "field" (DB field) to another field or to an object with the valorization of an attribute of the actual object. Likewise, it is important not to confuse the possibility to valorize an attribute "name" (denomination) or an attribute "code" (classification in the sense of encoding) of an object with regard to the class to which it belongs. A wall called "skirting board" will always be a wall for the machine.

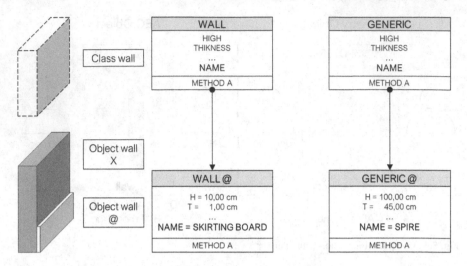

Fig. 5.4 Entity without class and generic class

This opens a further complexity in the use of BIM software (OOP with AEC ontology) integrated or inter-related with OOP software with other non-OOP ontologies or software.

One of the most interesting perspectives of BIM is the multidimensional approach of the "method," besides the 3D, in the integrated management of time (4D), costs (5D), useful life (6D) and sustainability (7D).

Technically this is feasible today owing to the relational potentialities of DBs, but with regard to BIM it is indeed far from being realized and realizable with the current software.

Let's take as object Wall @ (code: "*M.01*") with time t = 0, with attribute name: "*internal wall*," attribute type: "*plasterboard*" and attribute thickness: "*15.00 cm*." to which a defined graphic simulation of the BIM authoring software corresponds (Fig. 5.5).

The human operator or an algorithm predefined by the operator (in both cases without the possibility to control the standardized machine) links (through ID machine) to this object an item of estimative metric calculation belonging to a pre-structured electronic datasheet (cell 1: code "*C2.MW.001a*"; cell 2: "*concrete wall*"; cell 3: "*15.00 €*") and an item of work program belonging to a common Project Management software (field code: "*3.2.15*"; field WBS: "*plastering*"; duration field: "*25 dd*"). With regard to the calculation item, we do not know the thickness of the priced wall (15.00 represents Euros and not cm) and the material is "*concrete*" and not "*plasterboard*." Also with regard to the work program item, we do not know the thickness of the scheduled wall (25.00 represents the days and not cm) and the "*plastering*" work certainly is not an activity for "*plasterboard*" walls.

The machine, not understanding the meaning of the data entered in the cells and fields of calculation and work program, trusts the imposed link and (not having by

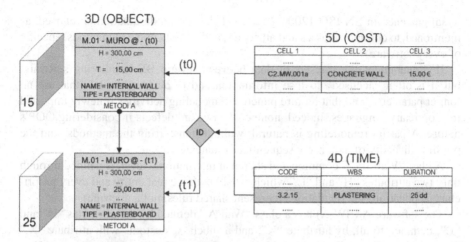

Fig. 5.5 3D, 4D and 5D software connection

nature uncertainties) it will continue to keep that trust (despite the initial errors and any following modifications). Not only does it follow the man's error but (in "its" precision) it reiterates it without solution of continuity.

Notice how each operator takes care of encoding and denominating the environment without this entailing any verifications or contraindications by the machine.

At time t = 1 the operator decides to modify the thickness of the Wall @ from 15.00 to 25.00 cm and, automatically, the BIM authoring software takes care of modifying the object, the graphic simulation, etc. What will happen to the calculation and to the schedule? Nothing. The other two types of software do not have an object Wall @, nor a valorized thickness attribute and even if there was a field or a cell this time identified as a place indicating "thickness," the machine would not be anyway able to interpret the data contained in it autonomously.

5.3 BIM Language: IFC

If, on the one hand, referring to real entities leaves several paths open for each software compared to the single methods with which to operate, on the other hand, it certainly reduces the variability (logic) of the attributes that can define each class. This at least makes possible, if not easier, to assume standards that allow the interoperability between numerous types of software operating on the same domain. In the AEC case, the international standard that defines its classes of reference is EN-ISO 16739:2017 known as IFC (Industry Foundation Classes) [10].

Therefore, IFC is BIM's open format (open BIM), written by Building Smart International and formalized, as mentioned, in the standard ISO 16739 (version 4.0:2013). In IFC, the classes and attributes recognized at international level are defined (the dominium is AEC and its ontology is expressed, even if with some misalignments, in EN-ISO 12006-2:2016) [11]. The standardization is referred, as mentioned, to common classes and attributes, and not, of course to methods (specific of every software).

IFC guarantees the transfer of data between several BIM authoring software, but it is often "accused" to lose information and to develop models that are no longer parametric and that require punctual remodeling activities if newly imported in proprietary languages. Indeed, none of the two are defects if considering OOP's nature. A partial remodeling is natural, without transferring the methods, and the possible diversity of data is consequential to standardization.

A class "Wall IFC," common and shared at international level, is defined through only two attributes "H" and "L," which satisfy both the real entity and every proprietary application (the regulations represent shared rules, not a doctrine).

The software A (SW-A) has a class "Wall A," defined by the attributes "H" and "L", common to all, by attribute "X", and a subclass "custom" with attribute "K." Generated an object "Wall @A," this will be istanced by the valorization of "H = 10," "L = 20," X = ertyz" and "K = 25."

Once the object "Wall @A" has been exported (not saved) from the software A in IFC (once defined its MDV, Model View Definition), we will have a "Wall @IFC," istanced through the standard attributes (IFCpropertySet) "H = 10," "L = 20." The methods, specific of each language, are not transferred (Method A ≠ Method IFC ≠ Method B).

The software B (SW-B) has a class "Wall B," defined by attributes "H" and "L" as well, common to all, and by attribute "Y" specific of B. Importing "Wall @IFC" can generate the object "Wall @B," defined by the sole standard attributes "H = 10," "L = 20," as there is no way to valorize "Y," both for IFC and for the original SW-A. Therefore, as evident, IFC has not lost any datum, as it transferred what is part of the shared standard (Fig. 5.6).

The unstandardized attributes can be anyway transferred as datum if exported as custom indication, not standardized, "IFCproxy." Even if, of course, a "proxy" datum has the property to be readable but not to be comprehensible (to be re-processed) by the machine.

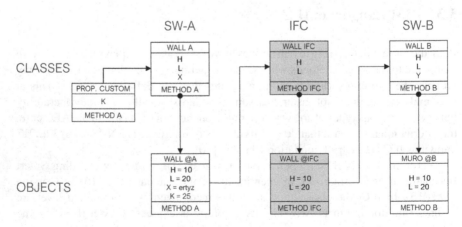

Fig. 5.6 Proprietary formats and IFC open formats

5.4   Obligatory Nature of Open Languages

With the introduction of digital modeling, a critical aspect newly came up, actually never solved, with reference to the guarantee of transparency, integrity and completeness of data and their transmission, regardless of the computer tool/language that produced them or that is managing and storing them.

This guarantee of completeness and integrity of the datum is obviously ensured by the software using proprietary languages, but only within the "proprietary" environment. Therefore, the proprietary language, in a dialogue with third parties, becomes a natural barrier to information transparency and transmission toward other types of software and computerized systems that use different languages. Software houses usually solve these critical aspects by keeping their own language, internally, and adopting systems for the transfer of data with an "open" nature tending outward. Conversion into open languages.

These languages are defined open because their structure is public (freely accessible). A language defined through a standard (ISO, EN, UNI, etc.) is considered "open" on the basis of its nature.

The open language guarantees the integrity of data transfer, but not always the full or the same operativeness if used through specific software different from the originating one. In the object oriented language, this involves the transferability of properties/attributes and the untransferability of methods (Fig. 5.6). Therefore, the operativeness on the data, generally speaking, takes place through a reconversion into proprietary language, or the use of specific software punctually modifying the file in open language.

A typical example of the use and typology of open languages is represented by word processing programs and the use of formats that are:

- open without composition rules such as *.txt;
- open with the maintenance of composition rules such as *.rtf;
- open with the guarantee of data integrity and composition such as *.pdf.

Each of the above is useful for a fundamental purpose, that is to guarantee the reading of the data—main aspect of open systems—but presenting obvious limits of use for other purposes. Another purpose of open formats is to guarantee the reading of the datum in its "original composition."

According to common practice with regard to word processing, the user sends a written document in proprietary format, e.g. "*.doc," or proprietary/open, such as "*.docx" (both of the Microsoft Office environment). The transferability, readability and changeability of the datum, as well as the structure of the document, are assumed on the basis of the dissemination in the market of the originating software (market standard, actual standard).

Rarely, instead, the user makes use of "open" formats, such as "*.txt" or "*.rtf" to guarantee transferability, readability and changeability (and with "rtf" also the structure of the document) regardless of the software used by the recipients.

On the other hand, always in the word processing environment, when needing to guarantee the reading and structure of the document, ensuring at the same time its integrity (also over time), the user modifies the previous attitude (careless of the receiving environment) and takes care of using an "open" format and in particular an open one—conceptually—*not changeable* ("not to be modified," if not with specific tools and in a much less efficient way). In this case, the user will take care of transferring the file as "*.pdf" (ISO 32000:2008; in its various variants "/A", "/X", "/E", etc.) [7].

Basically, the user delegates to the "*.pdf" format the guarantee to transfer information, but not the power/possibility to reprocess it. To transfer an "image" (static), the image of a thought, in quantitative and qualitative terms, stable. The main purpose to guarantee its reading is maintained, along with a second purpose, just as important, that is to guarantee the formal and substantial integrity (compared to the original) of the information to be transferred (Fig. 5.7).

In the CAD-AEC environment (Computer Aided Design—Architecture, Engineering and Construction) this critical aspect (content readability and integrity) in data transfer was overcome in the current practice by the dissemination in the market of a proprietary format (DWG, Autodesk) which has become in actual fact standard, representing the proprietary format mainly read and written (for market needs) by all other competing software. Of course, over the years, this has given a significant competitive advantage to the software house proprietary of that language, at the disadvantage of the remaining houses that have anyway created a market niche. At the same time, said houses have also had to guarantee their users the full compatibility with the dominant proprietary format (DWG) (Fig. 5.8).

As mentioned, with the introduction of BIM (Building Information Modeling), in the AEC environment, the same critical aspect occurred with CAD is now taking place. In order to avoid incomes from a job role (as occurred in CAD), which can limit

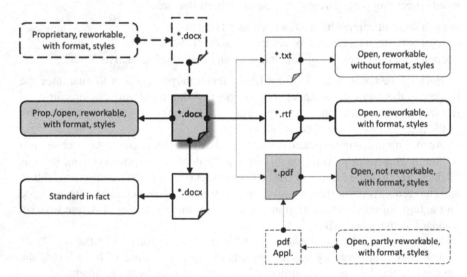

Fig. 5.7 Scheme of proprietary and open formats of word processing software

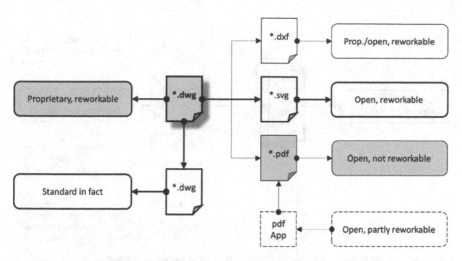

Fig. 5.8   Scheme of proprietary and open formats of CAD software

the progress and evolution of BIM, the problem of data transmission and reading, regardless of the software in use, was faced in this case from the very outset, before the dissemination of BIM in the markets. The solution identified involved the use of IFC open language (Industry Foundation Classes), promoted and supervised by the organization Building Smart International (former IAI, International Association for Interoperability, founded, among others, in 1985, also by Autodesk) (Fig. 5.9).

The IFC format, for the transmission of graphic models (UNI 11337-1:2017; information models produced through BIM authoring software), currently represents the only consolidated and "open" information certainty for the AEC sector.

The IFC format is an international standard (UNI-EN-ISO 16793:2016), based on the Express STEP standard (ISO STEP 10303-11:2004) [8]. It guarantees the "open"

Fig. 5.9   Structure of building smart international

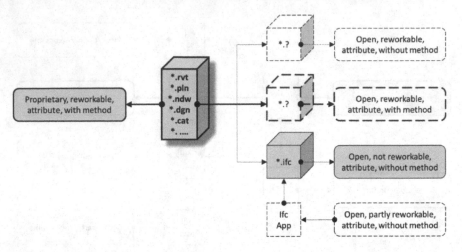

Fig. 5.10 Scheme of proprietary and open formats of BIM authoring software

reading of the geometries of objects (computerized) as no other language currently available. Information geometries and attributes connected to the same, according to IFC classes.

The guarantee of data transmission and storage over time is therefore assured through the use of unproprietary open formats, in BIM as in any other information system. With regard to BIM, today's most solid and widespread open language is the standard IFC format (Fig. 5.10).

It is indisputable that the use of open files is indispensable for the guarantee both of users and of the market. However, their current limits of real use lie in the fact of making such objective a mere matter of principle, as it has often occurred in the BIM environment toward IFC, without understanding and improving the existing limits, as instead it has been done for other open formats in other sectors. And, in the future, this may lead to their possible underuse, if not their abandonment.

If not adequately supported, updated or made performing toward old and new needs, the open format (IFC) will be overcome by the swiftness of the market (as already occurred in the past with CAD).

There has been a short-sightedness toward the users' needs, and more attention has been paid on imposing its use, by law, rather than fostering the efficiency and effectiveness of its actual use. However, the IT environment proves on a daily basis that using imposition in order to overcome the virality of use is not the right path. An example concerning the little attention paid toward the market is the fact that, to date (2019), there are no specific publications on its structure and modality of use (the most widespread is Autodesk) and there is only one certified software (BuildingSmart) for IFC 4 (although being an update of 2013) obliging users to operate "in actual fact" with the version IFC 2×3 (of 2008).

Year	Works	Cost
2019	Complex works	≥100 million/€
2020	Complex works	≥50 million/€
2021	Complex works	≥15 million/€
2022	Works	≥EU threshold
2023	Works	≥1 million/€
2025	Works	<1 million/€

Table 5.1 Mandatory introduction of BIM in Italian public contracts

Therefore, the problems, if existing (it would be more appropriate to talk about possible limits of use toward different uses considering its original nature: data transferability, guarantee of reading over time, implementation of clash detection between graphic models originated with different languages) are not related to IFC in itself, but at the most to how it is used, disseminated and scarcely innovated toward the daily needs of the market.

Today the Italian public administration, for example, imposes IFC in "BIM" contracts by decree (Lgs.D. 50/2016, Ministerial Decree 560/2017). At the same time, though, it often forgets to impose open formats for all other types of files, as if the graphic model were the sole type of document that will be produced in the contract. Moreover, in the schedule providing for the mandatory introduction of BIM, the first contracts planned (2019) are those relating to complex works above 100 million Euros. Very likely, big works such as railroads and highways. In IFC, though, specific classes for the infrastructure have not been defined yet (IFCRail and IFCRoads; Infrastructure agreements, MOUs Barcellona 2017) (Table 5.1).

Likewise, for example, in AEC contracts for works or services, contracting stations still often require for graphic models (rigorously in IFC open format) to be deposited, together with any other datum and file, in specific information management "Databases" managed by the contractor. However, contracting stations then forget to ask, with regard to said databases (DB): language, structure and architecture. Indeed, the latter are indispensible for managing the system after the deposit, and for integrating it in its beginning data structure (information system of the Contracting Station). Therefore, the "DB" necessary for BIM (CDE Common Data Environment BS 1192-1:2007 and PAS 1192-2:2013/ACDat, environment for data sharing; UNI 11337:2017) is confused as a simple file repository (an ftp—file transfer protocol— perhaps a bit more evolved).

Another aspect always neglected in the use of IFC is that it is not "saved" or "printed" (as a ".pdf") in IFC but it is "exported." What is exported and how, though, is just as important as defining the version of IFC (to date 2×3 or 4). Why does the "open" model need to be requested? In which phase and for what use and objective was it requested? Which attributes is it important to make sure are present also in the open model?

In other words, has the Model View Definitions (MVD) [5] of the open model requested been defined for the IFC version that will be used?[1]

To date, the IFC Model View Definition defined on BSI's website are [23]:

- IFC 4 Reference View;
- IFC 4 Design Transfer View;
- IFC 2 × 3 Coordination View (V2.0; V2.3);
- IFC 2 × 3 Structural Analysis;
- IFC 2 × 3 Basic FM HandOver view.

Since the rules and scopes of use of a PDF are clear, no one thinks that it is inadequate. Its purpose—be it big or small—results clear for the market and for users, and does not raise misunderstandings or false hopes. Today this is still not true for IFC. As mentioned, though, not as much for the tool in itself (improvable, as anything else, but already very effective), but because not much understood in its functioning (not very efficient due to its wrong use).

Given any two or more models, modeled in a proprietary language (e.g. architectural model—*.rvt—and structural model—*.ndw), these can be extracted in an open format (*.ifc; v. 2 × 3; MVD 2.0) to be shared between subjects without restrictions relating to software and modeling knowledge. Their purpose can be:

- static visualization (without any need of modifications) outside of the modeling software;
- deposit in CDE job order (shared or published) for storing and possible federation/aggregation;
- coordination for verifications related to clash detection and code checking (BIM review).

The different uses can be temporally and physically distinct or consequential and referred to a single collaboration and sharing environment, that is the CDE (Common Data Environment) (Fig. 5.11).

The deposit in CDE allows every subject to be acquainted with the information and production of other models (to work). BIM review operations allow to update and coordinate the existing ones (updating).

Therefore, the question is: Is the client and especially the public client aware, when rightfully asking for a "BIM" model in open format (IFC), of what this entails in terms of flow of consequent activities and of impact on procedures? It is an impact that must not limit the use of open formats, compared to the (illusory) convenience of the proprietary ones, but must be fully evaluated in terms of time, costs and resources to be used both for third parties (contractor) and for the Public Administration (contracting station). BIM cannot be imposed for its data transparency and then not be transparent in, or aware of, consequent processes in every aspect, also in open formats.

For example, two critical passages of the building process are [1]:

---

[1]The standard ISO 29481 in its part 3—that was supposed to define the rules for IFC's MDV—has not been published yet. This is another emblematic case concerning the fact that it is not IFC to have problems, but the inactivity in completing its instruments for its effective and efficient use.

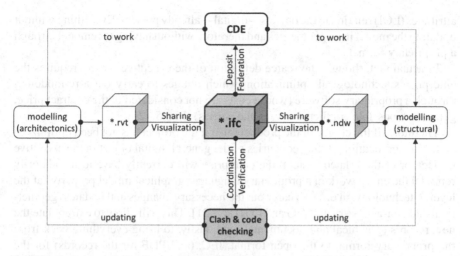

Fig. 5.11 Definition of objectives and uses of the graphic model in IFC open format

- the updating of the "executive" project model during the execution of the works (also without variants) up to the final production of the "as-built" model;
- the updating of the "as-built" model during execution (due to the natural lifecycle of the elements composing it) up to the production and storage of a "management" model (or of "execution").

For simplicity, let's consider the sole contract documentation connected to the graphic models: the digital records extrapolated from the models (excluding all other records) and the actual models.[2]

The model of the execution project, in the public contract, does not identify the brand and model of the composing products (Art. 68, par. 6, Lgs.D 50/2016) [13]. The enterprise, for its intervention (construction, recovery, etc.), receives the models (architecture, structures, MEP, etc.) from the Contracting Station in IFC format and the related graphics and documents, extrapolated from the proprietary models, in PDF format. In its constructive model, it will have to insert the product chosen for the type of work under contract on the basis of the requirements defined in the execution project. Therefore, these are not variants, but simple, natural improvements of the general data of the industrial product defined by the enterprise, in phase of execution, and accepted by the Contracting Station.

If an actual data sharing environment is available (CDE/ACDat, UNI 11337-5:2017) [22], and not a mere file repository; if the project is complete in all its parts and truly "executive;" and if the enterprise's choices only entailed the updating of the non-geometrical attributes (LOI) [21], the enterprise could act directly on the structure of the data connected to the IFC model and update the information

---

[2]With regard to the difference between graphic, documental, multimedia models and records see UNI11337-1:2017.

attributes (LOI) relating to the objects—digital—already present. Everything without updating the model in its geometries and therefore without having to remodel through a proprietary format.

In actual fact, though, the scarce definition of the executive project requires the enterprise's technological optimization, which obliges to carry out a re-modeling through a proprietary software (whose costs are not considered in the contract price, generating possible conflicts between the parties) (Fig. 5.12).

Therefore, if the choice of the product (also if the project has not been optimized) entails a modification of the geometries of the general digital object in the executive project, or of the related system, the enterprise will currently have to mandatorily remodel the entire work in a proprietary language (graphical model property) at the level of technology, in order to carry out the necessary changes and updates (geometrical and non-geometrical, LOG and LOI) [12, 21]. This will allow to extrapolate the new records (graphical and documental) and, lastly, to bring everything back from the proprietary format to the open format, IFC, (and PDF for the records) for the deposit of the As-built, at the end of the contract (in the test phase).

The geometrical change of only one or more components of the work obliges, in actual fact, to remodel the whole work in a proprietary language.

The enterprise's deposit of the As-Built in IFC guarantees to the PA data transparency, management and storage over time. However, the IFC model of AS-Built is not an update of the IFC execution model, but rather the remodeling through a construction model in proprietary format (Fig. 5.13).

On the other hand, as it already occurs for documents deposited in PDF open format, also IFC assures the enterprise of the occurred ("certain") deposit of a specific set of data. This example will not take into consideration the quality and quantity of the set of data, which depends, in a small measure, on the BIM authoring software used, and in large part, on the knowledge of IFC and of its MVD (Model View Definition), held by the operator that produced it.

Once the execution phase has been concluded, the As-built model is transferred, always in IFC open format, to the Site Manager and it becomes the beginning model for the execution phase (*management/execution model*) [9]. As the lifecycle of the work and of its composing elements advances, the manager maintains, repairs and substitutes the *"real"* objects and, at the same time, updates the attributes of the *"digital"* objects of the management model. This will mainly concern non-geometrical attributes (e.g. type of intervention, subject intervened, date of execution, cost, etc.). Since buildings are composed of elements/components with a useful life inferior to that of the actual building, they will be almost totally replaced over time, both in the real environment and in the model (the respective digital objects).

As already occurred, with regard to the *"execution"* and *"construction"* models (when passing from a general object to a production object), the replacement of the product installed in situ (irreparable breaking, end of lifecycle, etc.) with a new product that does not entail a mere updating of the non-geometrical information (replacements with an equal brand and type) can be indicated in the initial IFC model (as-built updated). Whereas, if this entails also a change of geometries (of brand or typology and, therefore, also of dimensions, shape, etc.), the Manager will

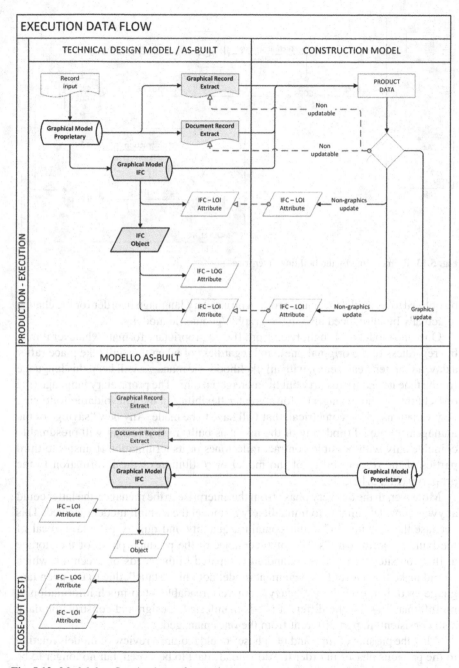

Fig. 5.12 Model data flow in phase of execution

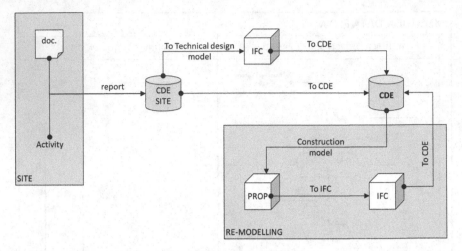

Fig. 5.13 Re-modeling by the building enterprise

be obliged to re-model the entire work in proprietary language in order for the change to actually be introduced also in the graphic/geometric model.

Once the model is brought back from IFC to proprietary format (whatever it may be, regardless of the original one), and regardless of when this will take place (after a day, a year, ten years, etc.), with all likelihood, the manager will keep this language for the time necessary to carry out his managerial duty. The proprietary language (the one chosen by the manager) allows greater flexibility in the unavoidable following new variations, also geometrical, that will have to be made. Any new "saving" of the management model (updating of the initial as-built) in IFC format will presumably coincide only with possible contract milestones or its termination (transfer to third parties of the responsibility of the model or restitution of the information to the property) (Fig. 5.14).

Moreover, in the delivery phase from the enterprise to the manager, the latter could anyway consider suitable to immediately re-model the as-built model received. This because the as-built IFC could contain a quantity and quality of data, typical of the building sector (the 's date of acceptance of the product, photo of the storage at the worksite, etc.) that is redundant compared to the needs of execution which would make the relevant management model not only "static", due to the open language used, but, especially, "heavy", not very readable—too much information = no information—for the different needs manifested. Design and construction data, also consistent, in part different from the ones managed.

Also, the passing of time and of "phase" could require a review of models relating to the previous phases in order to "download" data to be stored, but no longer to be kept active (Fig. 5.15).

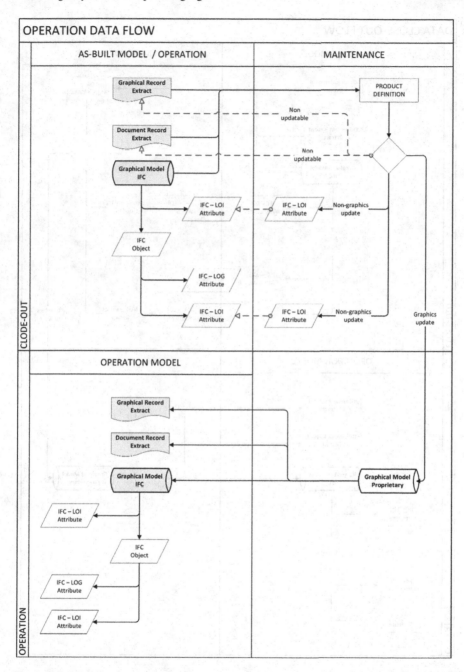

Fig. 5.14 Model data flow in phase of execution

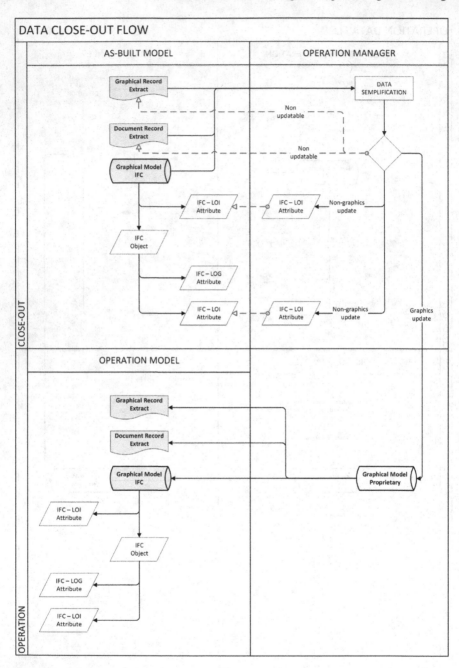

Fig. 5.15 Data flow of models in phase of delivery (enterprise-manager)

5.5   Standard Aspects: ISO 19650:2019 and UNI 11337:2017

With the introduction of IFC, a guaranteed collaboration of the BIM "method" can be ensured only through standards. As highlighted also by the EUBIM Task Group of the large community contracting stations (railways, infrastructure in general) "...*Without a standard data and process definition the supply chain and client will be re-creating a diverse range of proprietary approaches* [proprietaries BIM Guidelines] *which will potentially add a cost burden for each project...*" (EUBIM Task Group, "Handbook for the introduction of Building Information Modeling by the European Public Sector. Strategic action for construction sector performance: driving value, innovation and growth," http://www.eubim.eu/handbook/, par. 3.1.3, pag. 48) [4].

The panorama of the standards published at international, community and local level is by now very rich, starting from ISO STEP 10303, which gives origin to regulations relating more to IT aspects, up to the more recent process regulation ISO 19650-1-2:2018 (Fig. 5.16).

With the publication of the latest international standard (ISO 19650:2018), the general regulatory framework has been completed inserting the highest level process regulation that currently acts as a framework of principles of procedural nature from which the various specifications of more punctual nature derive. It has been implemented by Europe and consequently by every Member State (for Italy UNI EN ISO 19650:2019). The standard 19650 has only two national annexes to the standard group environment of PAS 1192 UK and the group of UNI 11337 (Fig. 5.17).

The structure of the regulations organized differently on the basis of themes and topics, can be schematized as follows with reference to the main international, local, UK, USA and ITA standards (Fig. 5.18).

Recalling the main concepts introduced by PAS 1192-2:2013 and 1192-3:2014 [20], the new EN ISO 19650-1-2:2019 [12] defines the general framework of BIM processes as follows:

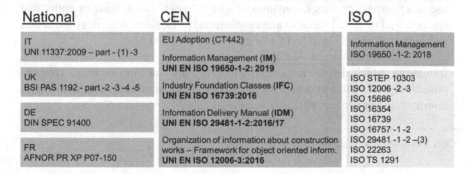

National	CEN	ISO
IT UNI 11337:2009 – part - (1) -3	EU Adoption (CT442) Information Management (**IM**) **UNI EN ISO 19650-1-2: 2019**	Information Management ISO 19650 -1-2: 2018
UK BSI PAS 1192 - part -2 -3 -4 -5	Industry Foundation Classes (**IFC**) **UNI EN ISO 16739:2016**	ISO STEP 10303 ISO 12006 -2 -3 ISO 15686
DE DIN SPEC 91400	Information Delivery Manual (**IDM**) **UNI EN ISO 29481-1-2:2016/17**	ISO 16354 ISO 16739 ISO 16757 -1 -2
FR AFNOR PR XP P07-150	Organization of information about construction works – Framework for object oriented inform. **UNI EN ISO 12006-3:2016**	ISO 29481 -1 -2 –(3) ISO 22263 ISO TS 1291

Fig. 5.16 BIM structure of international standard

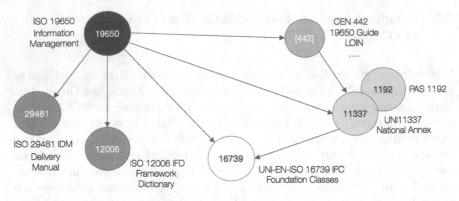

Fig. 5.17 Scheme of main international standards

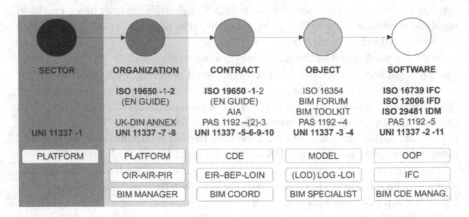

Fig. 5.18 International standards by arguments

– Building information modeling—BIM
  use of a shared digital representation of a built "asset" (item, thing or entity that
  has potential or actual value to an organization) to facilitate design, construction
  and operation processes to form a reliable basis for decisions;
– Level of information need—LOIN[3]
  framework which defines the extent and granularity of "information" (reinter-
  pretable representation of data in a formalized manner suitable for communication,
  interpretation or processing);

---

[3]CEN TC442 WG2N231:2019.

- Information model
  set of structured and unstructured "information container";
- Information container
  named persistent set of "information" retrievable from within a file, system or
  application storage hierarchy (including sub-directory, information file—including model, document, table, schedule, etc.);
- Appointing party/appointed lead and party
  receiver (client-owner-employer)/provider of information concerning work, goods
  or services;
- Exchange information requirements—EIR
  "information requirements" (specification for what, when, how and for whom
  "information" is to be produced) in relation to an "appointment" (agreed instruction
  for the provision of "information" concerning works, goods or services);
- Common data environment—CDE
  agreed source of "information" for any given project or "asset", for collecting,
  managing and disseminating each "information container" (named persistent set
  of "information" retrievable from within a file, system or application storage hierarchy) through a managed process;
- Bim execution plan—BEP
  plan that explains how the information management aspects of the "appointment"
  will be carried out by the delivery team.

Analyzing what consolidated over the years, also in the operational practices, the
most important changes introduced by ISO19650 are: the transit from *"Employer"*
to *"Exchange Information Requirement"* (EIR); the cancelling of the concept of
LOD (Level of Development—USA, Definition, Detail—UK) due to the introduction
of LOIN (Level of Information Need); the introduction of the Project Information
Requirement (PIR); the generalization of the concepts and deliveries of: Project
Implementation Plan (PIP), Master Information Delivery Plan (MIDP) and Task
Information Delivery Plan (TIDP), as sets of the programmatic activities of the
appointed (lead and party) of *Information Delivery Planning* (IDP):

- timing of information delivery;
- responsibility matrix:
- federation strategy and Information container Breakdown Structure (IcBS).

Basically, the concepts remained unvaried are those of Common Data Environment (CDE) and BIM execution Plan (BEP) (Fig. 5.19).

Since it is a procedural method regulation, ISO 19650 is very different from the
previous international regulations, more addressed to the regulation of computerized
functionalities and tools. Its definition is affected, of course, by the structure of PAS
1192 UK with a substantial contribution of the European states in their drawing up
and, at the same time, of the absence of the USA, especially in the final writing phase.

As mentioned, the reference is to management and process regulations (ISO
9000; ISO 55000; ISO 21500), with regard to which, before the definition of the
requirements (OIR, AIR, PIR, EIR), there is the need, not defined in ISO 19650, to

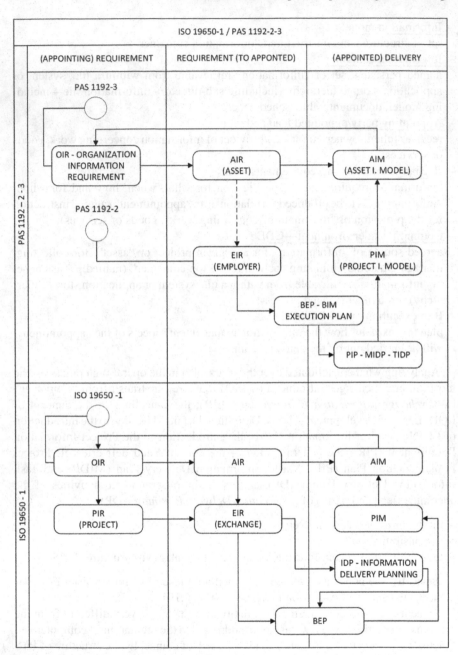

Fig. 5.19 PAS 1192–2-3 and ISO 19650 structure

establish a place for the consolidation of management rules relating to information flows at organizational level: BIM Information handBook (Organization-IB, Asset-IB, Project-IB), above those of a single job order: BEP (BIM Execution Plan) and IDP (Information Delivery Planning). It is also necessary to introduce the organization collaboration Platform that (with Organization Information Model—OIM), above all these, coordinates the data of the various CDE job orders (Asset CDE, with Asset Information Model—AIM; Project CDE, with Design Information Model—DIM, and Construction Information Model—CIM).

Both (contract) CDE and (Organization) Platform are defined in the Italian standard UNI11337 (ISO 19650 national annex) (Fig. 5.20).

Currently, there are only two national annexes to ISO 19650:2018 in the world. They are UK BS/PAS 1192, parts 1 and 2, and ITA UNI 11337, parts from 1 to 12 (national annexes to 19650-1 in the parts from 1 to 12, and to 19650-2 in the sole part 8).

UNI 11337 was published for the first time in 2009 as part 1 and in 2015 as part 3. To date, the parts available are 1, 3, 4, 5, 6 and 7, published from 2017 to 2018 (with translation also in English; part 1 was rewritten in 2017 replacing the original version of 2009).

Figure 5.21 shows the scheme of the Italian regulation, which is currently in complete review in the parts already published for a better alignment with UNI EN ISO 19650-1-2:2019, while parts 2, 8, 9, 10, 11 and 12 are being drawn up (in dark grey).

The Italian regulatory table on BIM refers to UNI's Construction Commission (033), which is composed, or was composed of 100 of the most important companies, bodies and organizations in the Italian building sector: public and private subjects, designers, producers, universities, research centers, software houses, associations, accreditation bodies, legal studios, publishers, etc.

The forming of the UNI table, since 2007, and especially its reopening in 2013, has allowed Italy to participate and cooperate with full rights in the drawing up of the most important regulations on BIM at international (ISO/TC59/SC13/WG13) and community (CEN/TC442/WG 01-07) level (Fig. 5.22).

Moreover, the Italian regulations define the IT differences between models and documents, also digital. They also introduce the concept of Collaboration Platform, the overcoming of the encoding systems (Tag.#BIM), LODs for restoration (historical buildings) and the worksite, the definition of rules for the automated verification of the administrative paperwork (@permit), and much more.

Key words of UNI 11337:2017:

- **digital collaboration platform**: a digital environment for an organized collection and sharing of data, information, models, objects and output for the building industry: resulting products, component products and processes (objects, parties, actions);
- **common data environment (CDE)**: an environment for an organized collection and sharing of related to digital models and output for a single work or a single complex of works;

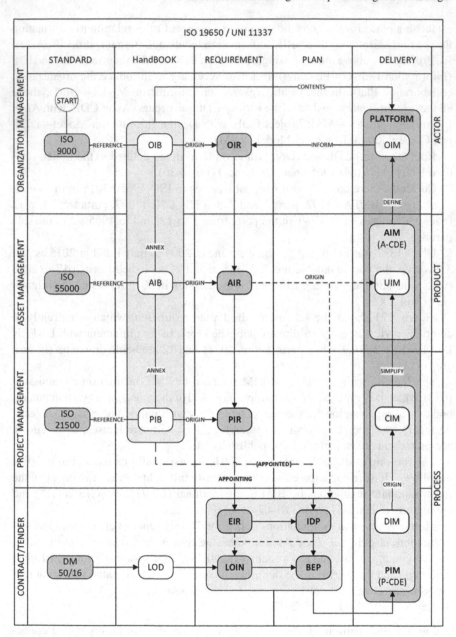

Fig. 5.20 ISO 19650/UNI 11337 information flow

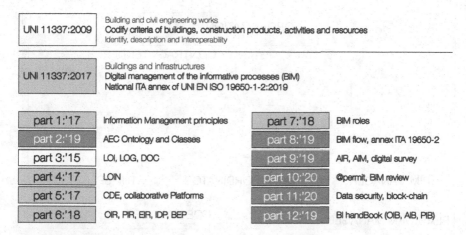

UNI 11337:2009	Building and civil engineering works Codify criteria of buildings, construction products, activities and resources Identify, description and interoperability

UNI 11337:2017	Buildings and infrastructures Digital management of the informative processes (BIM) National ITA annex of UNI EN ISO 19650-1-2:2019

part 1:'17	Information Management principles	part 7:'18	BIM roles
part 2:'19	AEC Ontology and Classes	part 8:'19	BIM flow, annex ITA 19650-2
part 3:'15	LOI, LOG, DOC	part 9:'19	AIR, AIM, digital survey
part 4:'17	LOIN	part 10:'20	@permit, BIM review
part 5:'17	CDE, collaborative Platforms	part 11:'20	Data security, block-chain
part 6:'18	OIR, PIR, EIR, IDP, BEP	part 12:'19	BI handBook (OIB, AIB, PIB)

Fig. 5.21 Italian standard UNI 11337 structure

Fig. 5.22 UNI construction commission standard 11337

– **object library**: a digital environment for an organized collection and sharing of object for graphical and alphanumeric models;
– **information content**: an information set organized according to a specific scope for a systematic communication a range of knowledge within a process;
– **information carrier**: a transmitting medium of information content
– **information output (output)**: an information carrier that represents building industry products and processes;
– **information model (model)**: an information carrier for virtualizing [simulating] building industry products and processes.

Moreover, first worldwide, UNI11337, in part 7 of 2018, clarifies the roles and functions of the new BIM figures, regulating the USA's and the UK's positions (BIM

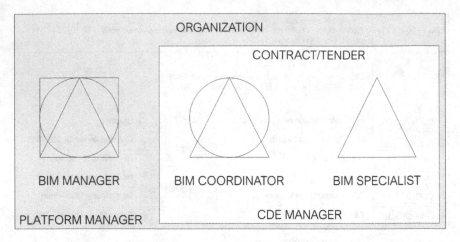

Fig. 5.23 UNI 11337-7:2018, BIM roles

manager/Information manager, BIM coordinator/Interface manager, BIM special-ist/information originator, etc.) taking the BIM manager at the level of organization and not of job order, introducing the figure of CDE Manager and opening the path to the future figure of Platform Manager (Fig. 5.23).

Among the highest international acknowledgments of the Italian regulations there are:

– the definition of the structure of BIM attributes for building products according to the EC Marking and the essential requirements of annex ZA (CPR 350/2011— Construction Product Regulation; UNI 11337-3:2015)
– the presidency of WG 2 at CEN 442 on the definition of European LOIN (Marzia Bolpagni, UNI expert from 11337) (Fig. 5.24).

5.6   From LOD to LOIN

With the introduction of CAD, the concept of design "scale" (1:100; 1:50; 1:25; etc.) was no longer used, having become a printing option (output) and no longer being a processing option (input). With BIM, the measurement system has been definitively closed with regard to information complexity and granularity mainly/exclusively connected to the graphic representation. The models highlighted the importance of an evaluation of the information quantity and quality (LOD: Level of Definition— UK/Development—USA) on the basis of the process phase expressed in graphi-cal terms (LOD; Level of Detail—UK/Element geometry—USA) or in ungraphi-cal terms—alphanumerical/multimedia (LOI; Level of Information—UK/Attribute information—USA).

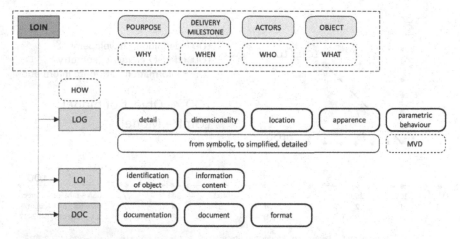

Fig. 5.24 Scheme of the Italian contribution for EU LOIN—UNI 11337 (A. Pavan)

The information measurement in BIM is therefore no longer defined on the basis of graphic scales, but of LOD. In other words, as a set of geometrical and ungeometrical information (LOD; LOI).

The most consolidated LOD measurement systems in the world still are (Fig. 5.26):

– the USA's system of the BIM Forum LOD Specification, relating to objects with scale from LOD 100 to 500;
– the UK's system of PAS 1192-2:2013 and of the NBS BIM Toolkit, the former relating to models, with scale from LOmD Brief to In Use, and the latter relating to objects, with scale from LOD 1 to 6;
– the Italian system of UNI 11337-4:2017, relating to objects with scale from LOD A to G.

With the publication of ISO 19560, the scheme of LODs that had consolidated over the years, even if with some differences among the various countries and regulatory systems, has in part been disrupted by the introduction of LOIN.

LODs used to define the gradualness of the information defined a priori (scales of LOD) to which to refer. LOINs (Level of Information Need), instead, open the information paradigm to a non defined/definable variability a priori because dependent on the specific needs of the moment (object, subject, phase, intervention, etc.) independent from each other (the same "product" in different phases, for different uses, in different environments, in a different timing, is subject to different information needs).

Moreover, the future European guide on LOINs of ISO 19650 (see Fig. 5.25) introduces the concept of "document" (DOC), data and information container, besides geometrical information (that become more clearly LOG) and non geometrical information (that remain LOI).

Therefore, LOINs are being defined (at least in CEN442) as:

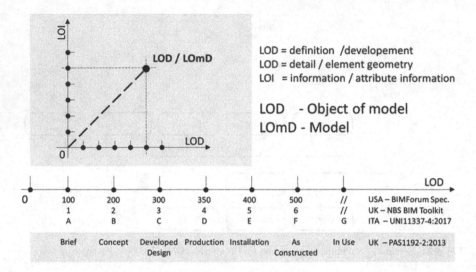

Fig. 5.25 Structure of LOD before ISO 19650

- LOG geometrical information;
- LOI non geometrical information;
- Documents.

Likewise, it appears evident that the "phases" and the phase objectives (strategy, project—in its various aspects, construction, execution, etc.) are defined/definable passages in the solidity of the datum (concept of "development" introduced by BIM-Forum USA) that would not be correct to neglect or cancel. Moreover, by introducing LOIN, the concept of data phase and solidity could still be assigned effectively to LOD.

At this point, with the introduction of LION, it is clear that it is also necessary to update the previous conceptual structure of LOD represented on the Cartesian plane (Fig. 5.26).

The new conceptual structure of LOIN could, therefore, be represented in the three dimensions as follows.

5.7   Coordination in Digital Public Contracts

At the end of the past millennium, the outsourcing of the Public Administration's activities became a recurring theme of economic policies also in economies strongly publicist, such as the European ones; this took place on the false line of strongly private systems such as the American one, through privatization policies of public or controlled/participated enterprises. To outsource "non-core" activities, concentrating resources only on "core" activities, is economically efficient and effective for any type of economic subject according to a theory by now widespread that modified

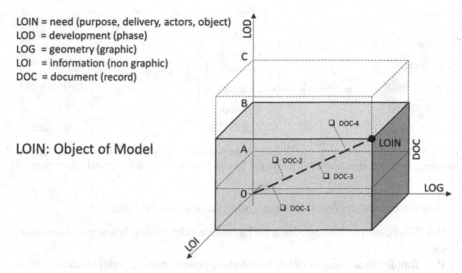

LOIN = need (purpose, delivery, actors, object)
LOD = development (phase)
LOG = geometry (graphic)
LOI = information (non graphic)
DOC = document (record)

LOIN: Object of Model

Fig. 5.26 Conceptual representation of LOIN

the paradigm of the multi-sector network acquisition typical of the corporate strategies of "multinationals" throughout the '900s (acquisition in correlated and similar sectors, and beyond, up to the most extreme diversification, from agriculture to the automotive).

In the building sector, the outsourcing of execution activities, first, and then of design activities, represents a historical theme faced and regulated already in the legislation at the end of the '800s for the execution of public works (Royal Decree no. 350, 1850).

Over the years, the P.A. in quality of Owner, Designer, Site Manager and Builder, first of all stopped being in charge of *Construction* activities, assigning them through contracts to specific external bodies, specifically dedicated to the different work activities involved, while it continued to maintain the roles as *Owner, Designer* and *Site Manager*.

Over the years, although maintaining the organizational role as *Owner*, the P.A. has started to outsource to third parties also the role of *Designer and Site Manager*. Initially, this was carried out if the workload exceeded the internal personnel or if of particular complexity; then, in a more systematic and continuous way, the P.S. started to outsource activities such as the *Services contract* (Lgs.D. no. 50, 2016).

Although outsourced, the roles as Designer and Site Manager are, by rule and in actual fact, considered dependent and symbiotic to that of Owner and P.A., such that the Works Contract and the Services Contract are viewed as formally and deeply different among each other in the relationship with the proposing subject: the P.A.[4]

---

[4]Figure 5.27 shows how the outsourcing of the roles as Designer and Site Manager never placed them in the common actions and thoughts outside of the P.A.'s scope, as instead it has always resulted evident toward the building sector and its outsourcing. Consultant designer, enterprise third party merely interested in the profit.

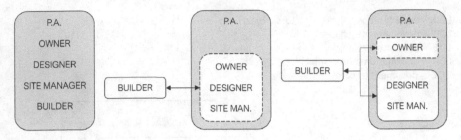

Fig. 5.27 Evolution of the P.A.'s outsourcing in the building sector (from left to right): sole internal functions; outsourcing of the building activity; outsourcing of owner, designer and site manager

The contractual approach in the common thoughts and actions:

– P.A./Designer-Site Manager, in a collaboration relationship (same common interest);
– P.A./Builder, in an antagonist relationship (opposed interests, public interest versus private interest).

Moreover, the current more consolidated contractual structure sees a diversification in the outsourcing also between Designer and Site Manager, with differentiated assignments and different subjects involved. Contrarily to what used to be normal with a consequential outsourcing of Designer and Site Manager to the same contractor.[5]

The latter differentiation represents a technically correct logic (Designer/Site Manager) but, considering its current organization, it has the intrinsic defect of further distancing the Builder (Enterprise)—that mainly dialogues only with the Site Manager (controller/manager)—from the Designer (deviser/innovator). Therefore, the execution activity is more and more "distant" (not contaminated) from the design activity.

The only difference to this scheme that has consolidated over the years, initially introduced as a solution for evident critical aspects related to management, in favor of an optimization of the outsourcing, and today instead unexpectedly denied (Lgs.D. no. 50, 2016), is represented by the introduction of the tool *Integrated Contract*. In this case, the P.A. in actual fact stopped being involved as Designer in favor of meetings for moments of devising and realization under the Builder and therefore under a sole subject in charge: the enterprise.

The P.A. instead remained in charge as (internal) Owner and Site Manager (then outsourced) (Fig. 5.28).

---

[5]Even relating to activities of very different nature: to devise an asset from scratch or its transformation on the one hand (inventiveness), to organize and control its correct construction on the other hand (management), tariff conformation (Law no. 143/49)—that penalized with an increase of expenditure the client that wanted to separate the two activities (+25%)—in actual fact blocked the choices towards the solution of the sole continuous more convenient assignment (project + Site Manager).

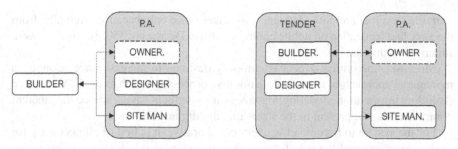

Fig. 5.28 Structure of P.A.'s outsourcing in the building sector, traditional contract of works (on the left) and integrated contract (on the right)

If the Integrated Contract was not the correct answer to balance the scarce managerial efficiency and the complicated effectiveness of the contractual relationships, also the consequent return to the previous (traditional) scheme, instead of a new in-depth rethinking of the system in itself, did not solve the basic problem.

To consider the Designer and the Site Manager as a privileged (consultancy) relationship with the P.A., although no longer with internal functions but outsourced, compared to the relationship with the Builder (mere supplier), places the P.A. in a difficult relationship, almost of subjection and certainly of untertiary, towards the various external subjects. With relations sometimes overestimated (Designers and Site Managers) on the one hand, and on the other hand often devalued, even before being underestimated (enterprises).

To consider the relationship with the Designer (or the Site Manager) different from that with the Builder, once they are both outsourced, has no economic or legal reason whatsoever, exposing the P.A. to a partial reading of the fact, filtered by a subject anyway "interested" (not properly third party) in the events. Sometimes, against its own interests. Moreover, the Designer (or Site Manager) is burdened with responsibilities that are not his due to the inability of management and cooperation/dialogue between the P.A. and the Enterprises. Designer and Site Manager are obliged to defend the P.A.'s non-fulfillments or deficiencies (thus weakening change and innovation) to then undergo the final consequences once in court.[6] The P.A., in this way, does not enhance the designer, but makes him captive, with the illusion of the fake alliance toward higher principles, that then clash with the crude reality of the Court rooms, to the disadvantage of professionals, called to pay for the P.A. (ethereal) and the enterprise (vanished).

In this way, therefore, the contractual relationships remain antagonist with the subject involved (outsourced), but they are (falsely) assumed to be collaborative

---

[6]One example for all: I do not pay an adequate survey and relevant evidence but then I give the fault to the designer and the enterprise for the obvious "unforeseen event" that could have been envisaged. Of course it is not a true unforeseen event, we all know that, but it is not so only if you budgeted it, and you make me do the survey and necessary tests, before the project. And who is supposed to solve everything, in everybody's interests, with works that cannot be stopped? The Site Manager that also used to be the Designer ....

with the Designer and Site Manager. This takes place only because originally, from the very first legislation on public contracts (Royal Decree no. 350, 1850), they were functions within the P.A.

Moreover, the critical aspect is currently destined to increase once adopting a managerial approach of strongly collaborative processes—such as the BIM oriented (Building Information Modeling/Management)—without having solved the inherent contradictions still present in the status quo highlighted above.

For the system to become effectively collaborative, it is first of all necessary for each subject external the P.A. to assume a role anyway subordinate toward it, but equal towards others and coordinated/referring with regard to the project. This is the sole common objective for all, in the specific timeframe planned. We are anyway talking about an economic activity, even if with public purposes, with the use of limited resources. It is not voluntary work or charity (Fig. 5.29).

At the same time, the P.A. must necessarily go back to playing its original role, truly third party and organizational/controller towards all.

Only in this way the external subjects' role can be collaborative among each other, in the mutual interests of achieving common project objectives towards the sole interlocutor external to the same (project), the owner (the P.A.). Moreover, this entails not only a joint involvement in the project, but also a concomitant involvement (temporality) with regard to the three functions: Builder, Designer and Site Manager so that the knowledge of the one is available for the other, not in a defined succession, but in progress (Fig. 5.30).[7]

In order for the system to be fully efficient and effective, also the Owner should assume a greater operational specificity (rationalization and professionalization of the Contracting Stations). This even up to the limit of outsourcing this function to professionals involved in real estate development and/or management (extending to

Fig. 5.29 Collaborative approach with the P.A.'s external functions

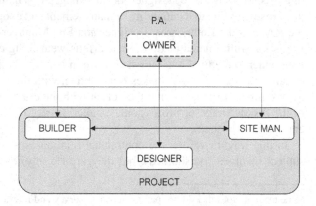

[7]In order to obtain the utmost synergy, Designer, Site Manager and Builder must collaborate with each other starting from the very devising of the project and not on the basis of a temporal succession. During the entire period of development (capex), up to the delivery of the asset to the manager for the execution phase (opex).

Fig. 5.30 Collaborative approach with all the operational functions external the P.A.

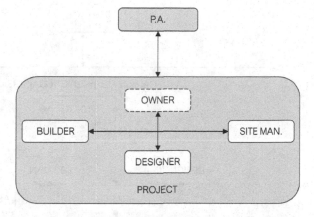

Fig. 5.31 Introduction of Asset Management function for operational phase

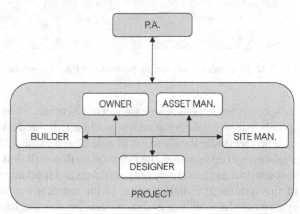

any possible work to be realized, as already implemented for the Granting or Public Private Partnership; considering the P.A. as user/lessor of its real estate) (Fig. 5.31).

As for any other "non-core" activity of the Public Administration, also constructions would assume in this way a full outsourcing organization, with the P.A. that would go back to assuming its leading role of regulator and controller, as well as following user (direct, instrumental spaces, or indirect, spaces open to the public).

Therefore, the project[8] is the economic place regulated by the P.A. where different subjects cooperate, professionalized third parties, having as their "core business" the execution of services and works in the building sector, with the aim to achieve a specific common objective, time and resources, toward a sole referring subject and controller (the P.A.).

These can act in antagonism among each other, resulting also co-responsible in achieving, or not, the common result (responsibilities and privileges).

Or they can fully collaborate with each other, (no one wins alone and everybody loses together), and anyway with the P.A., controller and regulator, not at all wisely

---

[8]From the Anglo-Saxon "Project" and not "Design".

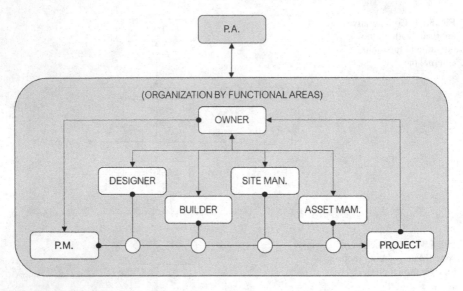

Fig. 5.32 Structure of outsourcing functions of P.A. in construction sector. Complete schema

antagonists towards their sole subject of reference. Therefore, the outsourcing corresponds to how a sole organization would act, divided into functional areas operating on a project under the direction of a dedicated Project Management (Fig. 5.32).

As in any organization, the information flow will thus become an aspect no longer secondary or even ancillary but overriding and fundamental for the good functioning of the system. The digitalization of the subjects becomes the actual bond for the synergy of the activities and resources, besides the driving force of collaboration. The same internal organizational structure of the subjects involved in the building process at various title will have to undergo a deep re-organization.

When the use of the tool will become a matter of fact ("BIM" tools as a natural accessory of a daily operativeness for all subjects, as it currently is for CAD and word processing), the overall information management will become a corporate function or a body in staff of the high management (Fig. 5.33) different from the traditional IT and from the mere use of need with regard to the use of the software (for this reason still today considered as an additional service that can be sub-contracted in outsourcing and not falling within the company's core business) [17].

After the serious facts occurred in Genoa (the fall of the Morandi bridge, August 2018), the Italian state decided to create a digital archive of public works (AINOP Archivio Informatico Nazionale delle Opere Pubbliche) and a single planning headquarters of all the public contracting stations. Of course, the use of BIM to define the two structures would have a very positive impact. Centralizing the direction of the planning, the information collection and the management at national level (as

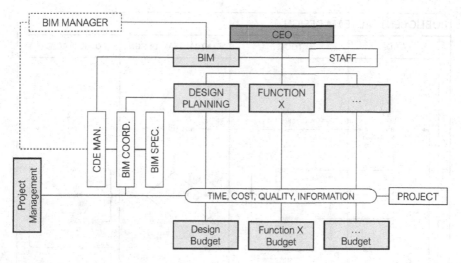

Fig. 5.33 Functional organizational chart of the organization

it already was in the INNOVance project), the prerogatives of collaboration and coordination of BIM can exalt the possibilities of success in achieving the results expected.

The following flow chart (Fig. 5.34) shows the possible future coordinated functioning of the two entities with the introduction of specific support tools, such as the CDE job order and information collection from the website (CDE Site), these too centralized as a support structure (cloud):

5.8 Conclusions

In the light of what mentioned above and confirming that it is fundamental to use open formats in order to guarantee data transparency, storage continuity and readability over time, and that it is fundamental for these (open formats) to evolve over time, according to the changeable needs of the market (because the datum, besides being transparent, has to be also efficient and effective so that the market may use it in procedures and not tempted to favor a proprietary format), it is possible to establish as follows:

1. it is necessary to define a stable and structured IFC ontology looking especially at the web environment as IFCOWL (overcoming the ambiguities of ISO 12006-2:2015 and the scarce orthodoxy of BuildingSMART Data Dictionary—bSDD);
2. it is necessary to widen IFC's classes at every real entity, also not merely constructive, necessary for the private and public processes (see for example the infrastructure: streets, railways, etc.);

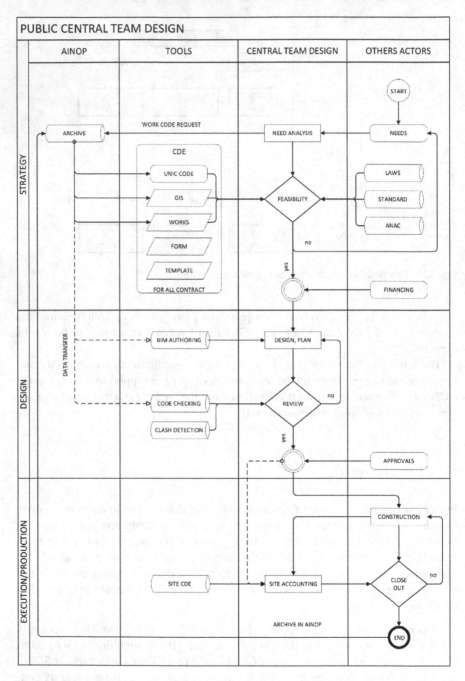

Fig. 5.34 Scheme of the functioning of AINOP and of the Sole Planning Headquarters of the PA

3. it is necessary to widen IFC's Property-set to each attribute necessary for the private contract and especially for the public one so that the completeness of the datum may be certain with regard to regulations and needs (limiting as much as possible the use of IFCProxy, if not as a collection of the evolutions in act on the market for language evolutions);

4. it is necessary to widen IFC's MVD so that they may be congruent with more objectives within the entire building process;

5. it is necessary to consider the current remodeling time and costs for some obliged passages of the process up to when new applications are developed capable of intervening directly on the IFC model with regard both to non-geometrical attributes and, especially, geometrical attributes;

6. it is necessary to define and collect beyond the graphical models—IFC but also proprietary—most of the non-geometrical attributes, so that their management and change does not have to mandatorily pass through the actual graphic models (ACDat as data model and not mere file repository);

7. it is necessary to develop software, api and platforms capable of managing and modifying the open formats directly and under the guarantee of transparency and visibility, especially the IFC format.

References

1. CEN/TC 395/WG 1 (2013) BS EN 16310:2013, Engineering services—terminology to describe engineering services for buildings, infrastructure and industrial facilities. BSI Standards Limited, London, p 36. https://bsol.bsigroup.com/Bibliographic/BibliographicInfoData/000000000030247589
2. Dahl O-J (2004) The birth of object orientation: the simula languages. In: From object-orientation to formal methods, pp 15–25. https://doi.org/10.1007/978-3-540-39993-3_3
3. Eastman CM (2018) Industry foundation classes. In: Building product models. CRC Press, pp 279–318. https://doi.org/10.1201/9781315138671-11
4. EU BIM Task Group (2016) Handbook for the introduction of Building Information Modelling by the European Public Sector. www.eubim.eu
5. Hietanen J (2006) IFC model view definition format. In: International alliance for interoperability
6. Ibrahim M, Krawczyk R(2003) CAD smart objects: potentials and limitations. In: Digital design: 21th eCAADe conference proceedings. Graz, pp 547–552. http://www.iit.edu/~krawczyk/miecad03.pdf
7. ISO/TC 171/SC2 (2008) ISO 32000-1:2008, Document management—portable document format—Part 1: PDF 1.7. International Organization for Standardization, Geneva. https://www.iso.org/standard/51502.html
8. ISO/TC 184/SC 4 (2004) 'ISO 10303-11:2004, Industrial automation systems and integration—product data representation and exchange—Part 11: description methods: the EXPRESS language reference manual. International Organization for Standardization, Geneva
9. ISO/TC 59/SC 13 (2008) ISO 22263:2008, Organization of information about construction works—framework for management of project information. International Organization for Standardization, Geneva

10. ISO/TC184/SC4 (2013) ISO 16739:2013, Industry Foundation Classes (IFC) for data sharing in the construction and facility management industries. International Organization for Standardization, Geneva
11. ISO/TC59/SC13/WG13 (2018) BSI 19650-2:2019, Organization and digitization of information about buildings and civil engineering works, including building information modelling (BIM)—information management using building information modelling. BSI Standards Limited, London, pp 1–46
12. ISO/TC59/SC13/WG13 (2019) BSI EN ISO 19650-1:2019. Organization and digitization of information about buildings and civil engineering works, including building information modelling (BIM)—Information management using building information modelling—PART 1: concepts and principles. BSI Standards Limited, London, pp 1–46
13. Italian_Government (2016) Dlgs no. 50, 2016. Italy. https://www.gazzettaufficiale.it/atto/serie_generale/caricaDettaglioAtto/originario?atto.dataPubblicazioneGazzetta=2016-04-19&atto.codiceRedazionale=16G00062
14. Kochhar S (1994) Object-oriented paradigms for graphical-object modeling in computer-aided design: a survey and analysis. In: Proceedings—graphics interface, pp 120–132
15. Laakso M, Kiviniemi A (2012) The IFC standard—a review of history, development, and standardization. J Inf Technol Constr 17(May):134–161. http://www.itcon.org/2012/9
16. Pauwels P, Zhang S, Lee YC (2017) Semantic web technologies in AEC industry: a literature overview. Autom Constr . https://doi.org/10.1016/j.autcon.2016.10.003
17. Pavan A, Mirarchi C, Giani M (2017) BIM : metodi e strumenti Progettare, costruire e gestire nell ' era digitale. I. Tecniche Nuove, Milano
18. Pauwelsa P, Sijie Zhangb Y-CL (2018) Semantic web technologies in AEC industry: a literature overview. Autom Constr J 95(1):185–203. https://doi.org/10.2139/ssrn.970643. ICE—Institution of Civil Engineers
19. Sutherland IE (2003) Sketchpad: a man-machine graphical communication system. Technical report. University of Cambridge Computer Laboratory, Cambridge. http://www.cl.cam.ac.uk/TechReports/
20. The British Standards Institution (2013) PAS 1192-2:2013, specification for information management for the capital/delivery phase of construction projects using building information modelling. British Standard Institute. British Standard Limited, UK. ISSN 9780580781360/BIM Task Group
21. UNI/CT033/WG05 (2017) UNI 11337-4:2017, building and civil engineering works—digital management of the informative process—Part 4: evolution and development of information within models, documents and objects. UNI, Milano, p 118
22. UNI/CT033/WG05 (2017) UNI11337-5:2017, building and civil engineering works—digital management of the informative process—Part 5: informative flows in the digital processes. UNI, Milano, p 24
23. Venugopal M et al (2012) Semantics of model views for information exchanges using the industry foundation class schema. Adv Eng Inform 26(2):411–428. https://doi.org/10.1016/j.aei.2012.01.005. Elsevier Ltd.

Chapter 6
Benefits and Challenges of BIM in Construction

Abstract This chapter covers the issues and benefits deriving from the introduction of digital processes and tools in enterprises in the building sector. A comparative analyzes is proposed between the current processes and the relevant information flows and the possibilities offered by the introduction of digital processes and tools. Starting from the different perspective given by the digital paradigm, the chapter analyzes how the increasing request for information and data and the need to produce information models that accompany the physical asset are changing the configuration of roles and relations between enterprises and the supply chain creating the need for new specialist management and collaboration structures (platforms) in the production phase. The second part of the chapter proposes a view on the possibilities offered by the introduction of machine learning systems for the management of information in enterprises. In particular, the potentialities of the current systems in organizing information and documents are analyzed for an improved management of the latter, both during the collection phase and with regard to the possibility to use the organization's historical documents in order to define the analysis processes.

6.1 Introduction

The digitalization of the building sector has led the sector's production paradigm to deteriorate, just as it consolidated during the past century.

Over the years, the division of the main subjects involved in the different phases of the production and management process of the real estate asset has increased: strategy/client, devising/designers, execution/enterprises and management/manager (Fig. 6.1).

The process, following its evolution, provides for a linear and consequential development of activities and responsibilities that evolve according to subsequent steps among: client, designers, general contractor and manager, with the direct intervention of the Public Administration, producers of components (manufacturers), specialist enterprises (and/or installers) and users (Fig. 6.2).

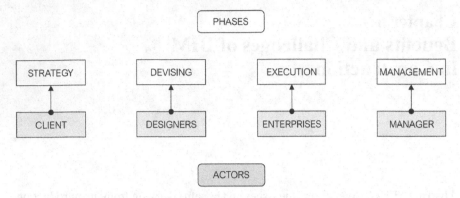

Fig. 6.1 Phases and actors of the building process

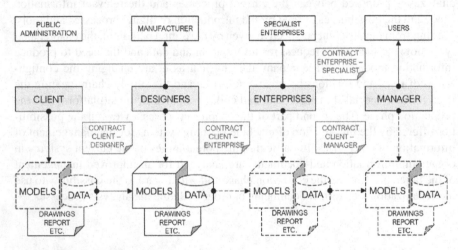

Fig. 6.2 Traditional information flow in the building process

In this system, the information flow is characterized by a qualitative/quantitative prevalence in the devising phase (project) preceded and followed by further moments of evolution, distinguished on the one hand in a temporal sense and on the other hand in a contractual sense, where each subject has a direct binding relationship only with the client and a merely antagonist relationship with the remaining subjects (with the exception of the enterprise that besides being under contract with the client, also has contracts with specialist/specialized enterprises). Each moment considers the preceding moment as defined and concluded and does not know or does not acquire anything of the following phase since no improving interaction is possible (beginning/end relationship; each step begins when the previous one is concluded, constituting a binding contractual milestone, see for example public contracts).

Differently, in the conceptualization of the digital information flow, a circular sharing system should be introduced not only with regard to information products:

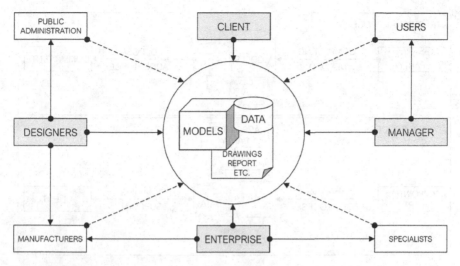

Fig. 6.3 Digital information flow in the building process

models, records, files, data, etc., but also to the actual "knowledge" of each subject involved, which is achieved without sharing contractual mediations, also temporal (everybody can act on the same databases), in a continuous circular development and no longer in a linear succession (Fig. 6.3).

The subjects directly involved remain the client, the designers, the general contractor and the manager, whose task is to dialogue with the public administration, the producers of components, the specialized enterprises, the installers and the users. All of whom could anyway have the need to interact directly for certain activities or needs, obligations, etc. (data extraction, signing of documents, etc.).

Therefore, from a linear procedural system such as the traditional one—with a mono-directional information flow carried out over time: from the client to the appointed party (out), and from the appointed party to the client (in)—the digital system should allow to transit to a circular procedural system, not only bidirectional but contemporaneous and collaborating, between client and appointed parties, and between each appointed party.

Therefore, the process involves passing from a system based on completing and depositing information with a counterpart subject (contractor—client), that acquires the documentation defined in every aspect (both specific aspects, partial deliveries/milestone, and as a whole, final delivery), to a system based on information sharing even if still in phase of production (not defined). In this case, each subject supports the process, even if client, and does not only receive the information produced by others (Fig. 6.4), but evaluates it (acceptance).

Moreover, the digital system results to be a process in which no one can consider oneself totally protected (or made responsible) from the current contract system, based on antagonist rules (of subjects) rather than collaboration rules (between subjects).

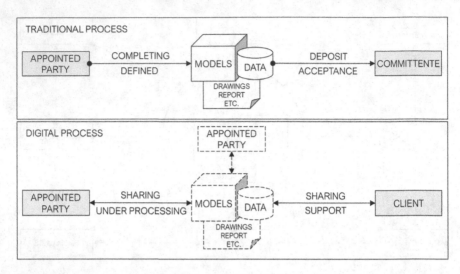

Fig. 6.4 Comparison between the traditional process and the process based on the digital system

Excluding the case of evident error, i.e. the sharing of information in progress, thus not defined and consolidated, this approach allows to support every subject involved for the maximum result possible on the one hand (collaboration of knowledge for a common objective), while on the other hand it currently exposes:

– who shares it, to following criticisms of inefficiency, incompleteness or inability;
– and those with whom it is shared, to higher and more careful responsibilities of analysis, verification and control (sharing of objectives, responsible for achieving results, etc.).

This leads to a necessarily hierarchical view of the environments of information exchange that must allow the subjects involved to develop activities and to collaborate as well. At the same time, it must facilitate the exchange of information according to models and rules shared between the different subjects involved clearly distinguishing between the moments of sharing (open) and those of development (reserved to the sole users directly involved in the actual development).

Digitalization introduced a system in the building sector according to which also the public or private client/producer (when not at the same time direct user/consumer) becomes the first subject responsible and of reference for the product and its performance towards the consumer, as it already occurs for every other industrial sector and of services.

It is important to highlight that also in the manufacturing sector the collaboration process is natural within the same company, between the different internal functional areas (marketing, sale, logistics, production, etc.). However, it involves every function otherwise outsourced only marginally, highlighting the same antagonist structure typical of the building sector: company/suppliers (manufacturing) and enterprise/specialists (building). Of course, the structure of the building sector makes the system even more complicated, since in general and in the typically traditional process it is based on: strategy, devising and production disconnected between each other and, the actual production, in turn newly divided between more subjects, always antagonists (general contractor, specialists, installers, etc.) as illustrated in Fig. 6.5. Moreover, these factors become complicated due to the spatial dislocation of the construction product (buildings, infrastructure, etc.) that make it a prototype, both as product and as production context/environment.

In this framework, therefore, also the enterprise's information flow (traditional/not digital) represents an underlying environment of the process further fragmented and antagonist towards its subcontractors (specialist enterprises), as illustrated in Fig. 6.6.

However, if also the digital flow is not dealt with according to a truly innovative approach, the enterprise's information relationship with the specialized enterprises and the installers risks to remain absolutely distinct, antagonist.

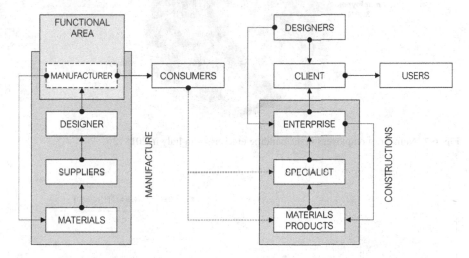

Fig. 6.5 Manufacturing and building sector

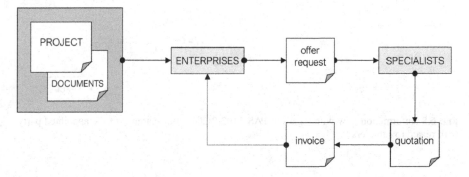

Fig. 6.6 Traditional information flow, building/specialist enterprise

The conformation of the sector—with a prevailing presence of small/micro enterprises (more than 90% with less than 9 employees; scarcely capitalized; not digitalized [2], Fig. 6.7)—also with regard to an information flow that has become digital, risks that neither the contractual relationship between subjects, nor the use and exchange of data and information will be modified.

The principle assumed in the British environment (PAS 1192-2:2013 [4], Fig. 6.8) regarding the absolute transparency of the general enterprise with regard to every

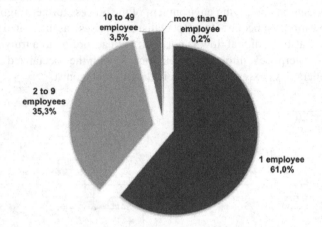

Fig. 6.7 Number of employees in the building enterprises in Italy in 2015 [2]

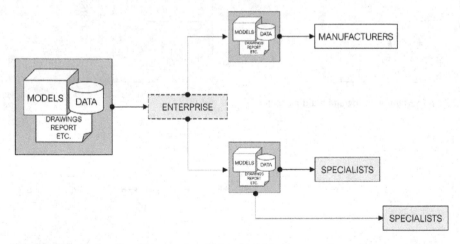

Fig. 6.8 Information flow with regard to PAS 1192-2:2013, transparency of the appointed party with regard to suppliers

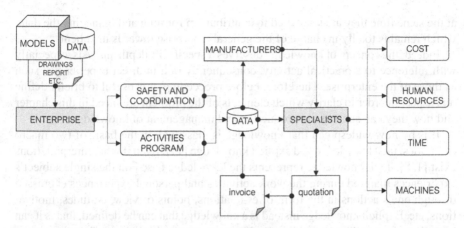

Fig. 6.9 Digital information flow in the production phase

subcontractor or supplier toward the information source remains currently impracticable, and not only for the national market[1] (contractual restrictions, requirements, models, etc. existing between the client and the enterprise are transferred and they are informed of every subcontractor so that they may conform). This is due to the clear difficulty for micro and small enterprises to satisfy structured requests of exchange of information, often including the use of complex tools and not much in line with the specialist activities developed by these realities.

With regard instead to the complexity of the digital flow, an information flow can be greatly assumable with the enterprise acting as filter and translator of such complexity towards subcontractors, in syntony with the companies that produce the components. And this with the addition, with regard to the project information received, of one's competences in terms of programming, safety management and coordination (Fig. 6.9).

The main position of the general contractor, obvious when knowing the market reality, allows a more efficient dialogue with producers and specialized enterprises and installers (specialists), but it still does not result to be effective with regard to the overall flow of knowledge. Part of the data, in fact, remains a prerogative of the subcontractor and does not fall within the information cycle if not under the usual static and "filtered" form (antagonist) of budgeting and final invoicing. The knowledge generated during the execution of the work by the entire work team falls apart at the end of the building activities, remaining confined in the sole minds of the professionals directly involved. Moreover, the lack of organization of the knowledge produced according to structured logics entails a loss of the context where the knowledge was generated, making its transfer even more complex.

In this configuration, the knowledge cycle is not completed. The various subcontractors are favored in managing the complexity of the digital information flow, but

---

[1] Also in England the building sector has a pulverized structure with more than 90% of the enterprises of the sector under 13 employees [5].

at the same time they are not called to contribute in forming and managing the data, which remains totally in charge of the general enterprise towards the client.

Indeed, the concept of knowledge deserves a specific in-depth analysis, especially with reference to a practical activity, consequently rich in direct experiences, such as those of the enterprise. Therefore, before proceeding, it is useful to briefly define this aspect in order to clarify which elements of the problem are faced in this chapter and how they can support the production and management of knowledge.

It is by now widespread that knowledge is classified on the basis of two macro categories, tacit knowledge and explicit knowledge [11]. Even if other interpretations exist [1, 8]. Tacit knowledge represents the knowledge based on the single subject's experiences matured during the work activities and personal experiences expressed through one's actions in the form of evaluations, points of view, attitudes, motivations, etc. Explicit knowledge instead is a knowledge that can be defined, that is it can be collected in handbooks, guides, databases, etc. On the basis of this first distinction, it is clear that explicit knowledge presents fewer critical aspects in its management compared to tacit knowledge, which instead is difficult to transfer since it is based on direct experiences occurred in a specific context. For this reason, the transfer of tacit knowledge often passes through the narration of stories and anecdotes that try to recreate the context where the knowledge was formed making the other users re-experience the event. On the other hand, explicit knowledge can be organized in a systematic way also owing to the support of computerized tools that can optimize its organization. The work carried out by Nonaka and Takeuchi [11] identifies in this sense four modalities for the transfer of knowledge, that is socialization, externalization, combination and internalization (SECI model—Fig. 6.10) creating a matrix that combines the two macro-classes of knowledge, thus enhancing the creation and transformation cycle of knowledge from tacit to explicit and vice versa.

Actually in this model the combination is often viewed as a not-real transfer in terms of knowledge in its direct form, that is in the direct transit from explicit knowledge to other explicit knowledge. In fact, in order to be generated and/or modified, knowledge needs to be acquired and integrated in the mind of the single subject that

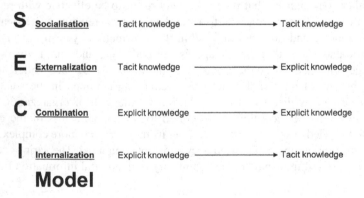

S <u>Socialisation</u>   Tacit knowledge   ⟶ Tacit knowledge

E <u>Externalization</u>   Tacit knowledge   ⟶ Explicit knowledge

C <u>Combination</u>   Explicit knowledge   ⟶ Explicit knowledge

I <u>Internalization</u>   Explicit knowledge   ⟶ Tacit knowledge

Model

Fig. 6.10 Representation of the SECI model

can then redistribute it explicitly. Therefore, also in the case of combination, it is indispensible to identify an intermediate transit, that is a configuration such as "explicit knowledge—tacit knowledge—explicit knowledge." Likewise, with regard to the collection of data, information and explicit knowledge, that represent the focus of this chapter, it is necessary to take into consideration aspects relating to the organization of knowledge and its integration in the mind of those that will be using it [6, 13]. It is therefore fundamental to identify tools and processes that can be adapted and that can define and form a base of shared knowledge in terms of semantics and ontology, as well as a specialist customization that allows the single subject to obtain and evaluate only the information interesting for the same on the basis of the context of interpretation that said subject considers as the best for him. The attention will thus be focused only (and necessarily) on a portion of the entire knowledge produced during the building activities, being well aware of the importance of the direct and personal knowledge of the single subjects involved (tacit knowledge). Consequently, corrective actions with regard to management will have to be necessarily implemented in order to limit its falling apart at the end of every project. The introduction of machine learning techniques and artificial intelligence briefly discussed further on, are widening the possibilities offered in the generation of knowledge starting from the large quantity of data, making their systematic collection increasingly important and of value, through which it is possible to develop the necessary background for the application of such technologies.

6.1.1 Collaboration Platform Within the Sector

In order to balance the support provided to micro-realities and the necessary collection of operational data essential to complete the information cycle towards the client and the other actors involved, it is necessary to assume a collaboration platform within the sector that simplifies data management and standardization in order to optimize the flow without burdening the consequent duties and bureaucracy (Fig. 6.11).

Against the collection, simplified through the use of specific devices and tools, of the operational data of the specialists' knowledge patrimony, the enterprise assures, through the collaboration platform within the sector, specific services that for micro-realities can include also invoicing and accounting management, difficult to access and often entrusted to not very effective unprofessional/manual processes.

Therefore, every subcontractor, in the new digital paradigm, is called to contribute toward the production of the data necessary to complete the digital process in the same way in which said subcontractor is called to provide work and/or products for the building process. Nevertheless, in an extremely fragmented and under-capitalized sector, the new digital information flow must not generate additional costs but, on the contrary, it must be used effectively to promote the quality of the final product (buildings, infrastructure, etc.).

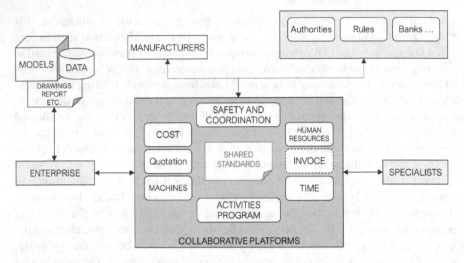

Fig. 6.11 Digital collaboration platform within the sector

This chapter analyzes the building enterprise's point of view on the framework of relationships and information flows typical of the relationship between the general contractor and their supply chain.

6.2 The Building Enterprises' Context for the Implementation of BIM

The enterprise plays a central role in the building process with its consequent centrality also with regard to the entire chain that follows in this process. In fact, on the one hand, it relates with the requiring and planning part of the process that thus includes clients, professionals and administrations; on the other hand, it relates with who will then use the product, such as the supply chain that, along with the enterprise, will realize the work. Such relationships can be divided into two macro-categories, that is internal and external relationships (Fig. 6.12). The former refer to the relationships with the enterprise's supply chain and with the producers whose relationships can be governed directly by the general enterprise. The latter refer instead to the relationship with the other actors involved in the process whose roles and consequently collaboration relationships are not under the enterprise's direct control.

The relationships between these subjects, as well as their role within the process, are progressively changing on the basis of the new possibilities and the new requirements imposed by the introduction of the digital paradigm. The supply chain—traditionally called to supply products, services, work, etc., that is to physically produce the work—is currently more and more involved in the production and collection of the information necessary to develop digital models that are useful for the planning

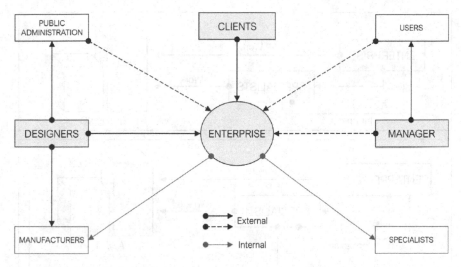

Fig. 6.12 Framework of the relationships in the enterprise's perspective

and construction phase, but also (and especially) for the future phases of management and maintenance of the real estate. The management of the information flows
from the enterprise to the supply chain and, especially, from the supply chain to the
enterprise, is of fundamental importance today in order to guarantee not only the
quality of the asset produced and the efficiency of the production process, but also
the quality of the digital models that will accompany the product during its lifecycle.
The sub-suppliers should therefore integrate themselves in a clear and well structured
flow in order to guarantee an easy exchange of information in the various phases in
which they are called to be involved (Fig. 6.13).

   In this framework, the definition of stable relationships between general enterprises and the supply chain could result of great benefit, enabling an incremental
improvement of data exchange practices and the consequent progressive definition
of protocols shared between the various subjects. However, the peculiarities of the
sector impose strong limits on this perspective on the basis of two main characteristics. On the one hand, the configuration of the projects dislocated on the territory
(each work must be realized directly on the territory with a consequent spatial fragmentation) limits the definition of stable and structured supply relationships since it
is often indispensible to recur to ad hoc supply contracts on the basis of where the
works are carried out (beyond a certain distance it becomes anti-economical to use
retained supplies, especially considering the massive components of the work—think
for example of concrete). On the other hand, the business structure of the supply chain
is often extremely not homogeneous and can include very small realities that do not
have the means, the competences and often the interest to use complex computerized
tools, as well as large and extremely structured manufacturing enterprises that for
this reason have well defined internal standard processes and tools that undoubtedly
cannot be adapted and/or modified on the basis of the single enterprise.

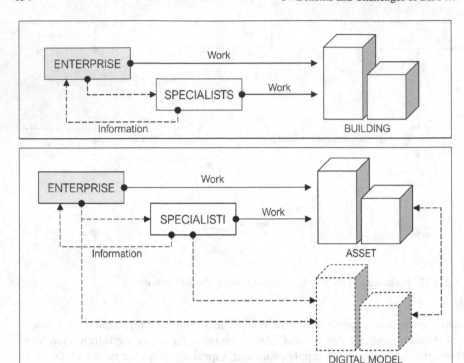

Fig. 6.13 Evolution of the role of the supply chain in the production process

Besides the supply chain, the enterprise has to deal with the project part of the process from which the fundamental information is drawn to realize the work on the basis of the client's requirements, materialized by the professionals. The net separation among the process phases eliminates de facto the possibility to define standards and practices shared between the actors involved, generally reducing the effectiveness of the transmission of the information.

The enterprise thus has to operate in an extremely structured context where on the one hand it receives project information without being able to control it, and on the other hand it has to translate this information in order to transmit it to its supply chain so that the latter may be able to efficiently meet the requirements. Moreover, it is called to collect the information generated during the entire production process and transfer it in an information model that will have to be functional for the maintenance activities and the management of the work.

It is important to highlight that the enterprise, although being de facto the producer of the asset, is traditionally excluded from its management and maintenance activities in complete contrast compared to the other industrial sectors. The definition of digital processes could change this configuration introducing new business models in the sector, on the basis of the increased ability of the enterprise to preserve and exploit the knowledge generated during the realization of an asset. This would allow to acquire new market activities providing at the same time services useful for the maintenance of the functional quality of the buildings during their lifecycle (currently representing an extremely critical aspect of the built patrimony on the basis of a progressive loss of performance in terms of efficiency, comfort and safety with major impacts both in environmental and social terms).

6.2.1 The Relationship Between Enterprise and Supply Chain

Focusing the attention on the enterprise and on its supply chain, it is possible to recognize two main moments in the exchange of information. First of all, the general contractor must transmit the information of a project to the specialist enterprise (filtering the information of interest for the latter—at least partially) in order to receive an offer (quotation) relating to the supply of materials, services, etc. The specific supplier will answer this request with its offer possibly integrating it on the basis of its fund of knowledge. Given the fragmentation of the sector and the continuous change of the subjects involved in the supply chain (on the basis of the peculiarities mentioned above), the general contractor often refers to more than one supplier for the same supply, so as to choose the one that is more fit for the specific intervention. Each one of these tends to provide offers according to personal standards or, often, without following any at all. This burdens the general enterprise with the responsibility to interpret and compare the various types of information received in order to select the best proposal, to then integrate it in its definitive version within the general offer to be submitted to the client (Fig. 6.14).

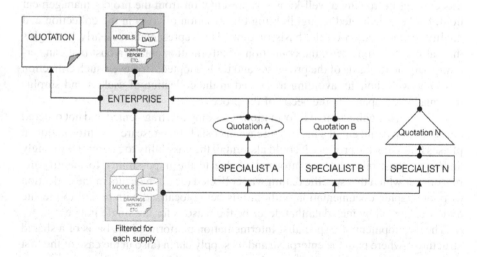

Fig. 6.14 Relationship between the general enterprise and specialist enterprises in the offer phase

Fig. 6.15 Qualitative trend
of costs relating to
information collection and
information quality on the
basis of time

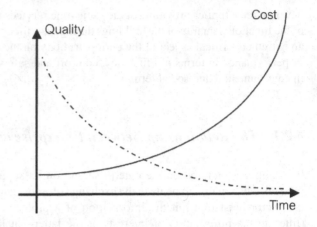
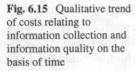

Secondly, once selected the components of the supply chain, these have to transmit the production information from the worksite to the general enterprise (Fig. 6.13). Given the great dynamicity of the production process and the concentration of subjects that can operate at the same time on the same project, it is fundamental to manage the information flow effectively and swiftly. The late collection and/or non-collection of the information coming from the production site often entails an exponential increase in costs for finding the information up to, in some cases, the impossibility to find the information searched (think, for example, of the collocation of concrete reinforcements if these have not been written down and/or photographed). Moreover, the information collected in the mind of the operators and/or collected in a fragmented way, while waiting to be transferred, quickly lose their quality. Figure 6.15 represents this concept borrowing a well-known representation from the project management field, but here dedicated to highlighting the variation of costs in the collection and quality of information on the basis of time. This representation clearly highlights that, also with a time zero, the collection of information entails a cost that can certainly vary on the basis of the processes and tools adopted. However, such collection has to be well thought, avoiding to exceed in the collection of useless and surplus information compared to the needs of the process.

Likewise, the collection of information according to a fragmented and not ordered configuration could entail a major increase in costs for the research of information in phases following the process. It could also entail the possibility to interpret it wrongly and/or the possibility to lose information due to the impossibility to identify the reasons for which that specific datum was collected (as often occurs in the collection of photographic documentation without this being localized with regard to the site and/or enhanced by metadata that describe the reasons for the single image).

The development of a specialist intermediation platform on the basis of a shared structure—where both the enterprise and its supply chain (also in the case of the first assignment) are capable of effectively communicating all the information necessary—could greatly increase the performances currently registered in the production phases.

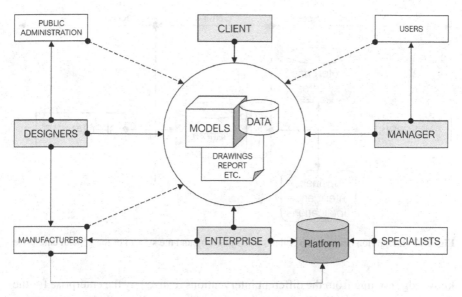

Fig. 6.16 Optimization of the digital information flow in the building process

The digital information cycle can therefore be optimized as follows (Fig. 6.16) with regard to the sector and the network of general and specialized enterprises and their sub-contracted installers.

It is evident that such environment will be characterized by a high complexity deriving from the richness of the problems to solve; a complexity that must not manifest itself on the users of the system that, on the contrary, will have to deal with extremely simple and straightforward tools and interfaces capable of providing answers in the time and ways requested from the production environment.

The processes defined through the platform will have to produce a tangible advantage also for the small supplier that has to recognize first of all the need to modify its practices and correctly integrate these processes. In fact, it could be counterproductive to impose processes without these being understood by who has to implement them, because said subject could use them without awareness and competence. The collection of "dirty" data could entail a worsening of the processes underway, requiring from the enterprise a further work of cleansing and reprocessing of the data to integrate them in the required models.

6.3 Integrate Digital Data in a Business Context

The digital collaboration platform within the sector, as described above, does not solve in itself the problems of data management and organization within the general enterprise in a view of business intelligence. That is, the reuse of the patrimony of

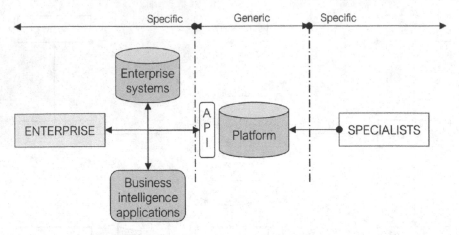

Fig. 6.17 Integration between the collaboration platform and the system of the general enterprise

knowledge coming from the different interventions realized by the enterprise for the development of decisional processes based on data as well as for the optimization of processes on the basis of the past knowledge. In fact, if on the one hand the collaboration platform could turn out to be a useful manager of all the information collected, including that relating to the financial and economic management for micro-realties, this appears not very functional in the case of general enterprises often already provided with structured systems for the integrated management of their data (Enterprise Resource Planning—ERP). Therefore, there is the need to investigate the modalities for a dialogue and integration between the collaboration platform and the tools and processes already used in the general enterprise. The collaboration environment can therefore be structured according to two macro-levels. First of all, the platform will have to be capable of dialoguing and transferring information according to a shared logic at the system level within the building sector. In this sense, the platform will thus have a "general" component (Fig. 6.17) capable of creating a common base of dialogue that can therefore be used regardless of the definition of structured relationships with the supply chain. This structure will have to be accompanied, though, by the possibility to specify/specialize the general components of the platform in order to adapt to the structure of the single subjects and satisfy specific needs, thus identifying the second macro-level.

On the one hand, this factor could be solved with regard to micro-enterprises through the immediate customization of the interfaces interacting with the platform (also on the basis of the specific project), so as to facilitate the interaction between user and tool.

On the other hand, the general enterprise could need more structured relationships with the data collected in the platform, both on the basis of the integration issues mentioned above and due to the need to integrate the information received in specific information models for the single project. Therefore, there is the need to develop specific interfaces of API communication (Application Programming Interface) that

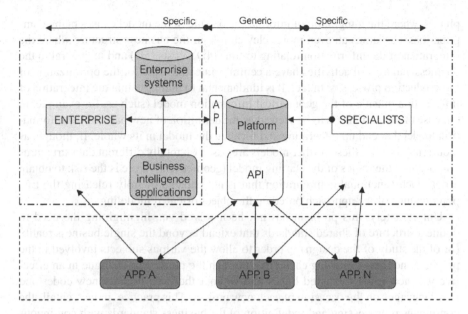

Fig. 6.18 Development of dedicated services based on the collaboration platform

allow a single enterprise to develop ad hoc applications capable of capturing the information in an automated way and transfer it to its management systems (Fig. 6.17).

On the other hand, the same principle could be applied to the sector in general, opening the doors to the development of services dedicated to constructions by the IT sector, as well identified in the INNOVance project and made available (also if with different logics) by the French platform Kroqui [12].

The digital collaboration platform could therefore provide a shared structure of data accessible and integrable through specialist applications, developed both by the single realities of the building sector and by the other sectors, increasing the level of digitalization within the entire industry (Fig. 6.18).

However, the deep discrepancy between the structure for objects inherent the BIM information models and a relational structure with different logics typical of the most widespread systems used within the companies, makes the configuration of this dialogue extremely challenging. If, on the one hand, the building information models are composed of objects belonging to specific classes and therefore governed by rules and parameters according to a shared semantics, this is not true for most of the business management tools. Therefore, the integration of the information collected between existing systems and information models deriving from BIM processes requires de facto a management that is still rather unprofessional based on an encoding often incapable of embracing the complex structure of the sector and therefore subject to continuous adjustments that invalidate the repeatability on different interventions. This situation—actually already manifested in the planning phases, where the integration of information outside of the model causes many complications—assumes an even more relevant role in the business structure where the integration of the information relating to time (4D), costs (5D) and in general to the business managerial activities have a central value and logic for the optimization of the production processes. In fact, it is fundamental to highlight that the integration of information outside of the geometrical information model (such as, for example, in the case of 4D and 5D) is to be viewed as the definition of new parts of the information model that end up integrating and forming the model in its whole. If, though, as it currently occurs, these further models are based on totally different data structures compared to the logics of the starting model (geometrical), there is the risk to obtain a not much functional configuration that is not capable of fully releasing the true potentialities of the approach based on the object oriented modeling.

Unfortunately, this problem has not been solved yet, highlighting the need to define a structure of shared standards that extend beyond the single business reality (or of the study of the design) in order to allow the various subjects involved in the process (and their following each other through the phases) to dialogue in an effective way according to shared logics and without the need to learn new codes and new languages on the occasion of each new project. This practice, in fact, entails the continuous reprocessing and redefinition of the business standards with continuous manual activities and consequent errors in the coordination between the different types of information. This makes the process fall in the usual errors that have long led to the development of project records with contrasting and not coordinated information.

Reasoning systemically on the theme of the platform, it is clear that what is shared is the structure of the data, that is the model according to which the data are organized, and not the actual data that will be reserved to the sole organizations directly involved. Nonetheless, the possibility to statistically collect information coming from a multiplicity of enterprises could greatly increase the analysis ability of a sector with a pulverized nature and that has difficulty in reaching a critical dimension of volume of data capable of releasing the potentialities offered by the most recent analysis applications.

Even if focused on the optimization of the collaboration in the project, the collaboration platform is therefore a basis for the integration of a knowledge otherwise fragmented and dispersed along the building process, and even more in the transit from a project to another, on the basis of the disruption of the relationship between enterprise and supply chain.

6.3.1 Automated Management System of the Information Flow

The introduction of machine learning (ML) techniques could change the scenario described up to here on the basis of a growing adaptability of the machine. ML constitutes a field of artificial intelligence that currently embraces a multitude of applications including image recognition, the reading and classification of texts in a natural language, the analysis of frauds, etc.

Besides the "obvious" applications of data analysis driven by ML, as already underway in other sectors, the framework of relationships described up to now outlines a path that could turn out to be of extreme value for the use of these techniques. In fact, if the transfer of information between the various subjects involved in the building process is currently governed manually and requires the use of standards for its optimization, the ability of the machine to handle the information and redistribute it according to logics similar to the human ones could modify the perspective of the analysis.

Let's imagine, for example, the process that allows the enterprise to translate the information on the project in order to be able to distribute it in an ordered way according to the supply chain. This operation, often managed manually, could be managed by the machine, optimizing the process and without any need to require standard data in phase of collection (Fig. 6.19).

The same principle could be applied also to recover information during the execution phases, or better in their effective organization within the business structure, so as to facilitate its use and reuse in the various activities. For example, the emblematic case of the images collected during the execution of the work and collected in block without specific references, could be solved (at least in part) though classification and/or clustering techniques capable of aggregating similar images, so as to create more effective research structures improving the accessibility of the information.

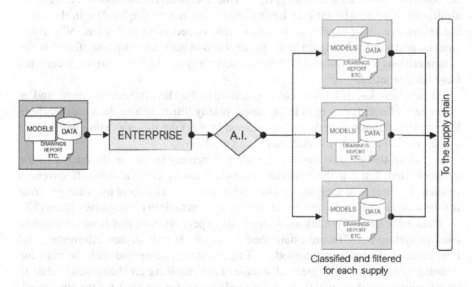

Classified and filtered for each supply

Fig. 6.19 Automatic sorting of models, data and records toward the supply chain through artificial intelligence systems

The ML techniques are mainly divided into two large classes, "supervised learn-ing" and "unsupervised learning" [7]. To put it simply, the application of supervised learning techniques provides for two distinct moments of development of the system. First of all, it is necessary to instruct the machine in the specific activity required. Once instructed, the machine will be able to act "autonomously" in the execution of the activity learned. To give an example of such process, let's think of the automatic classification of a set of documents. Starting from the collection of documents (obvi-ously in digital format readable by the machine) it is necessary first of all to identify the possible classes (or categories) in which the specific group of documents could be divided. Once defined such categories, it is necessary to manually classify a part of the documents available, so as to create a series of homogeneous groups of docu-ments classified correctly. This activity allows to transfer the operator's knowledge to the machine. By defining specific evaluation criteria, the machine can analyze the documents classified with the aim to identify peculiarities and recurring patterns that can identify with clarity the belonging or non-belonging of a document to a specific class (Fig. 6.20, upper part). Once the learning and evaluation of the results is con-cluded, the machine will be ready to classify new documents never seen before on the basis of predefined classes without the operator's intervention (Fig. 6.20 lower part).

In the case instead of unsupervised learning, the process does not need an initial training, but it self-regulates itself on the basis of the information received during the execution. Referring to the organization of documents, the machine starts from a set of documents comparing the information contained in said documents on the basis of predefined rules. It then aggregates the documents in clusters that can collect similar documents without ever having seen them before (Fig. 6.21). In this case, the knowledge of the experts of the sector (always necessary in this kind of activity) enters a posteriori with regard to the process and in particular it is manifested in the denomination (or identification) of themes according to which the various documents have been grouped.

These examples, that can be easily extended to the classification of images and/or of portions of documents or of information, briefly illustrate how the use of machine leaning techniques can make up for (at least in part) the limits existing with regard to information contained in documents and models.

It is clear, though, that these systems are extremely subject to the quality of the data collected, that is to the input data received. Consequently, also the effectiveness of such systems has to keep into consideration the current limits of tools that generate information models and apply due corrective actions to limit the negative effects [10].

Machine learning systems work alongside expert systems, that is an information analysis system based on pre-established rules that can provide new information and thus enhance the information models. These systems, compared with the machine learning systems, have the great advantage of not requiring great amounts of data to be effective (something which instead is fundamental for the ML). On the other hand, their effectiveness depends on the semantic structure on which they are developed and

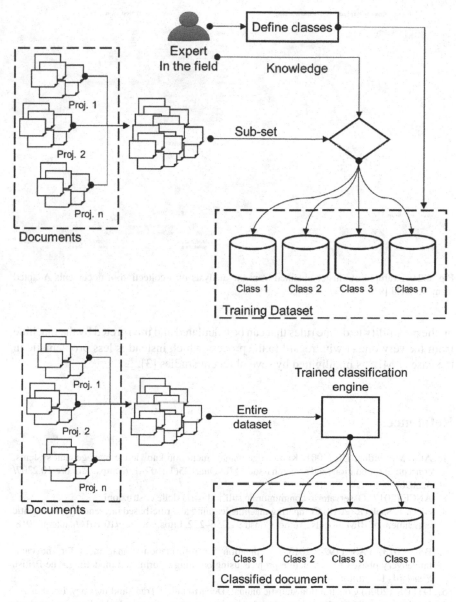

Fig. 6.20 Training and execution processes in the case of Supervised learning applied to the classification of documents Adapted from Mirarchi [9]

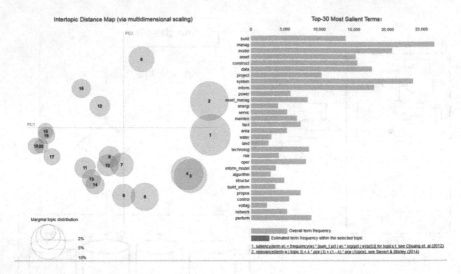

Fig. 6.21 Example of cluster obtained from the analysis of a collection of documents Adapted from Mirarchi [9]

on the possibility to define rules that can be calculated and interpreted by the machine from the very outset with regard to the process which instead is less problematic in the case of ML, as highlighted by several recent studies [3].

References

1. Alavi M, Leidner DE (2001) Knowledge management and knowledge management systems: conceptual foundations and research issues. MIS Quart 25(1):107–136. https://doi.org/10.2307/3250961
2. ANCE (2017) Osservatorio congiunturale sull'industria delle costruzioni
3. Bloch T, Sacks R (2018) Comparing machine learning and rule-based inferencing for semantic enrichment of BIM models. Autom Constr 91:256–272. https://doi.org/10.1016/j.autcon.2018.03.018
4. BSI (2013) PAS 1192-2:2013, specification for information management for the capital/delivery phase of construction projects using building information modelling. The British Standards Institution, UK
5. DTI UK (2006) Construction statistic annual. Department of Trade and Industry, London
6. Fernie S et al (2001) Learning across business sectors: context, embeddedness and conceptual chasms. In: 17th Annual ARCOM conference. University of Salford, 5–7 Sept 2001, pp 557–565. http://www.reading.edu.my/web/FILES/innovativeconstructionresearchcentre/icrc-18-i-ARCOM.pdf
7. Lecun Y, Bengio Y, Hinton G (2015) Deep learning. Nature 521(7553):436–444
8. Lundvall BÅ, Johnson B (1994) The learning economy. J Ind Stud 1:23–42
9. Mirarchi C (2019) Knowledge network for innovation of construction sector: increasing efficiency through process digitisation of the entire chain. Politecnico di Milano. https://www.politesi.polimi.it/handle/10589/145773

10. Mirarchi C, Pavan A (2019) Building information models are dirty. In: 2019 European conference on computing in construction. Chania, Greece
11. Nonaka I, Takeuchi H (1995) The knowledge-creating company: how Japanese companies create the dynamics of innovation. Oxford University Press, New York and Oxford
12. République Francaise (2018) Kroqui platform. https://www.kroqi.fr/. Accessed on 26 Sep 2018
13. Tversky A, Kahneman D (1973) Availability: a heuristic for judging frequency and probability. Cognit Psychol 5(2):207–232. https://doi.org/10.1016/0010-0285(73)90033-9

182 Bailliez

Chapter 7
Benefits and Challenges Using BIM for Operation and Maintenance

Abstract Considering the remarkable shift that the digitalisation is nowadays bringing about in the building sector, the chapter presents how data and information collected and managed during design and construction stages improve building operation and maintenance. In particular, the chapter focuses on how the great amount of dynamic data collected around assets during the operational stage is changing the way buildings are experienced and managed. The integration and sharing of information supported by collaborative environments and recent information technologies enhance the management of the built asset. Within that context, the chapter outlines benefits and challenges in adopting BIM-based processes for the operation and maintenance of buildings. Particularly, the chapter presents how an ordered and structured information management allows delivering buildings as service providers, extracting knowledge from real-time data for tracking user behaviours and designing user interactions with buildings. The results allow: (1) implementing workflows for enriching building information in the operational stage and, consequently, operating buildings with an increased value originated by information. (2) Assessing how buildings work in the operational stage, especially taking into consideration the influence of users. (3) Defining strategies for engaging different actors in building operations and informing them about the behaviours of both buildings and users. (4) Providing control strategies when unexpected behaviours (e.g., energy-hungry behaviours, unusual comfort conditions and FM-related failures) are registered. Considering the concept of Industry 4.0, also the collection, storage and fruition of data collected in real-time is considered for an improved building operation and maintenance.

7.1 Introduction

Different kinds of information are collected during building operation. Among that great amount of information, the chapter mainly presents how those data can be used for considering user behaviours, focusing on data collected around buildings through sensors. Indeed, nowadays, buildings are equipped with several devices for monitoring their use and performance, also tracking user behaviours. These devices

© Springer Nature Switzerland AG 2020   167
B. Daniotti et al., *BIM-Based Collaborative Building Process
Management*, Springer Tracts in Civil Engineering,
https://doi.org/10.1007/978-3-030-32889-4_7

monitor, among the others, indoor conditions, outdoor weather, room occupancy or user preferences. Through these devices great amounts of data are collected, with a great potential for changing the way buildings are not only designed and constructed, but also experienced and operated. However, these data are barely used for an improved management of buildings during their operation and, especially, these data are barely used for tracking the real behaviours and conditions of both users and buildings. Hence, limited advantages are generally derived by these data as several barriers still exist for making data available and valuable in real-time to users. This is mainly due to two issues. On one side, users with their daily activities greatly affect the behaviours and the performances of buildings they occupy, even if they are not aware of their influence. On the other side, these data cannot be accessed, interpreted and utilised by users because of the poor quality of those data and because of the lack of easy-to-use tools for accessing, analysing and processing data. Moreover, difficulties are still experienced for managing data and information that are dynamic and are produced autonomously and routinely by various forms of sensors with high sampling rates.

Hence, in order to derive benefits from data collected around buildings during operational stages, several barriers affecting the building process have still to be overcome. Among these barriers, it is worth to mention:

- lack of user engagement in building operation and lack of awareness on information potentialities (being users responsible of 30% waste of energy because of their behaviours);
- inefficiency in information management within traditional building processes, also due to interoperability-related issues in sharing information (costing 4.5% of construction turnover and considering that more than half of costs of inadequate interoperability is referred to obstacles encountered in operation);
- lack of tools for managing dynamic information;
- smooth delivery of data and information to the operational stage, also considering the limited implementation of BIM-based processes in operation.

Therefore, for overcoming outlined barriers, an improved management of information collected in operation can rely on the definition of information requirements, the adoption of effective solutions for improving information flows, the exchange of information ensuring interoperability, the identification of proper model uses and connected information to be collected and the adoption of BIM-based solutions for operation.

7.2 Defining Information Requirements

Many benefits are offered by the implementation of BIM along the building lifecycle. However, its adoption for facility operations is limited because of several barriers. Among those barriers, difficulties are often encountered by owners in identifying and formalizing the information requirements needed to support model-based project delivery and asset management [3].

Therefore, for an effective and efficient adoption of BIM, it is significant to develop and formulate BIM requirements to support the lifecycle of owners' assets.

Within that context, it is important to collect asset owner requirements [3]. Those requirements are generally grouped and classified as organisational requirements, project requirements, personnel requirements and BIM requirements [3].

Organisational requirements represent owners' general requirements about built infrastructure and space, system design, equipment and component selection, design and installation.

Project requirements refer to project specific requirements that are in more detail than the owner organisation's requirements.

Personnel requirements are informal and undocumented O&M personnel requirements for design, information, and information interaction, and access.

BIM requirements are listed as guidelines for using BIM in an organisation [3].

7.3 Adopting Information Flows

Several benefits derive from ordered and uninterrupted information flows. Indeed, data and information collected and managed during design and construction stages improve building operation and maintenance. However, typical design and construction information has only very limited content for O&M purposes [12]. Moreover, uncertainty in using data derives from incomplete building documentation [11].

Furthermore, data can be stored in an as-built model of buildings not only considering data that come from previous stages of the building process (i.e. design and construction), but there is the possibility to collect data gathered from different sources. Particularly, considering operational stages of buildings, data can be the result of on-site energy audit or on-site data gathering.

However, among the main challenges to be considered for an improved information flow, it should be mentioned the importance of declaring contractual obligation to deliver the necessary data, avoiding the transmission of superfluous information or information without a specific use or purpose [21]. Indeed, an overload of information causes a lack of purpose, transforming information in unused data.

7.4 Ensuring Interoperability

For sharing the content of BIM along the building lifecycle, interoperability needs to be guaranteed both among actors and tools. Particularly, interoperability among information systems concerns the need to transfer data among applications, preventing re-creation or re-input of data and enabling efficient use of information. Problems in interoperability are mainly due to redundant data entry, redundant IT systems and

IT staff, inefficient business processes, and delays indirectly resulting from those inefficiencies [6]. That means software non-interoperability costs on average 3.1% of total project budgets [22].

In order to overcome those barriers, different standardised formats and data structure specifications have been developed for sharing information as IFC, ifcXML, COBie and gbXML. Moreover, solutions based on semantic web are increasingly adopted.

7.4.1   IFC

With respect to interoperability-related issues, buildingSMART defined a structure to support an efficient exchange of information through Industry Foundation Classes (IFC), International Framework for Dictionaries (IFD) and Information Delivery Manual (IDM).

IFC specifies how information is to be exchanged, being a standardised neutral data format to describe, exchange and share information. It is an object-oriented data model of buildings, specifying physical items or abstract ideas and their relationships (decomposing, assigning, connecting, associating, and defining).

7.4.2   COBie

COBie (Construction Operations Building Information Exchange) is a format for delivering assets' data. COBie is a data format for the publication of a subset of information focused on delivering building data, not geometric modelling. COBie is a spreadsheet being considered as an IFC model view or as a subset of IFC with the necessary information for the management of a building. Indeed, COBie is mainly adopted for management; however, it is often composed of uncompleted spreadsheets, without any kind of information, reducing its use for effective building operation.

7.4.3   gbXML

For transferring building information from BIM authoring platforms to environmental (energy in particular) analysis software, gbXML (Green Building eXtensible Markup Language) is mostly used as an open data scheme. It is not registered as official standard, and it is applied only to energy analysis. A basic comparison between IFC and gbXML is conducted in terms of their data structure, representation and application (Table 7.1). With a comprehensive top-down data schema, IFC shows

Table 7.1 Criticalities in using information for O&M purposes

Information availability and accessibility	[7]
Stakeholder capability	[7]
O&M process effectiveness	[7]
Technology, process and organisational challenges	[3]
Project based business versus lifecycle management	[3]
Complexity of the implementation process	[3]
Difficulty in understanding the effective use of information for daily work processes	[3]
Owners are not aware of the complete set of information they require to support asset lifecycle information	[3]
Not enough experience for exchanging and managing information through BIM	[3]
Owners unsure about how to require information	[3]
Management of data that are dynamic and out of date	[15]

potential benefits in its highly organised and relational data representation. In contrast, the bottom-up gbXML schema is simpler (as it is based on XML language) and easier to understand, which facilitates quicker implementation of schema extension for different design purposes. Hence, the gbXML schema is easier to implement, but does not include all building (and its elements) information, as it is mainly targeted towards energy simulations. Furthermore, gbXML captures information representation, but not the relationships among them. The IFC schema, more complex and including larger scope of building aspects, is not yet common in sustainability analysis software, due to the difficulty of its implementation. However, as the aim of the research is not only to manage data gathered through sensors for environmental analysis, proposed digitally-enabled workflows have been based on the adoption of IFC (when the reference to open standards is needed).

7.4.4 Semantic Web

Furthermore, issues concerning semantically-related data interoperability are outlined. First, the information must be represented by computerized codes in a programming language. Second, ontology is required to define the relationships among concepts of building information-coded, computerized language [8]. To use BIM-based data relationships in a semantic inference procedure, IFC files have been converted to Web ontology language (OWL) for input into Semantic Web applications [11].

7.5 Identifying Model Uses

Once information is correctly and orderly stored, different uses of that information can be set. Considering the importance of setting a specific purpose for the use of stored information [1], several researches have been performed demonstrating how information collected during operation can be adopted for FM-enabled BIM [19].

Indeed, information can be adopted for preventive maintenance, by associating information about duration or scheduling needed intervention for a proper use of a specific system. Therefore, information can be used for scheduling of facility maintenance work orders [2].

Moreover, by comparing data collected about equipment and appliances, it is possible to define specific intervention when unwanted values are registered. In that case, undesired values can be the cause of equipment failures so that specific and immediate intervention can be defined for solving issues connected to registered values.

Furthermore, by collecting data about building use, it is possible to constantly control building conditions. The adoption of specific ICT solutions allow to adapt building condition to desired values, e.g. for setting comfort condition.

In all the presented case, a great support is provided by the application of data mining techniques [13, 18] and through the implementation of Internet of Things [4].

Indeed, data mining and machine learning can be used for improving the quality of collected information in order to provide facility managers with spatio-temporal visualization of the work order categories across a series of buildings to help prioritize and streamline operations and maintenance task assignment [13].

By coupling data mining and BIM implementation, the value of information can be increased for FM purposes, by finding relationships of similarity among records (cluster analysis), by detecting input improper data and keeping the database fresh (outlier detection), and finding deeper logic links among records (improved pattern mining algorithm) [18].

Moreover, coupling BIM processes and IoT paradigms, advantages deriving from a dynamic BIM use can be derived with respect to satisfaction of comfort conditions, space management and equipment failures [4, 15].

With a specific focus on data collected in operation through sensors, in order to outline the influence of users on the way building function for reducing the performance gap, user behaviours affecting building conditions can be detected on real case studies so to define customised strategies considering real users' preferences and patterns. Within that context, after cleaning data for excluding outliers or missing values, data can be analysed on platforms for highlighting trends and relations between behaviours. After that, those data can be used for feeding simulations [17].

Particularly, concerning an information-driven asset management relying on a user-centric approach, the chapter outlines the influence of user behaviours on building conditions during operational stages. The presented solutions rely on the definition of control strategies for engaging end-users in building operation while addressing behavioural changes. For reaching the declared purpose, it is important to focus on:

- information structure to be defined for collecting information, supporting the definition of information contents required for achieving specific model uses;
- information redundancy and interoperability affecting conventional building processes, analysing different information exchange processes for avoiding data losses, relying on proprietary and open solutions and formats and on ad hoc methods;
- information transmission among different sources, introducing workflows for managing great amounts of data and information that are dynamic and produced with high sample rates;
- information fruition on behalf of different users, proposing solutions to make information available and valuable even when it is stored in an ordered database, through BIModels, web-based interfaces, apps, tools for supporting augmented reality.

7.6 Adopting BIM-Based Solutions for Operation

For making information available to users in the form of control strategies, there is a need for digitally-enabled workflows for exchanging and updating information, connecting different data sources, relying on the development of both open and proprietary solutions.

Different aspects have been considered in the development of the technological solutions, as:

- the cost of manufacture, distribution and maintenance (e.g. comparing web-based applications to dedicated display);
- the rapidity of system updating;
- the ease of access by different suppliers;
- the possibility to control information from different places;
- the opportunity to send processed data to customers.

Within that context, the research couples the issues of managing great amounts of data gathered through ICT solutions with the concept of BIM-based processes, considering BIM-based solutions as a means for managing dynamic data and information gathered through different sources or devices. Indeed, the dynamic use of data is ensured, e.g. with reference to BIM models, where stored information is accessible not only through BIM authoring platforms, but also through web interfaces and mobile applications, thanks to the integration of BIM processes and IoT paradigms.

A structured framework has been outlined for the creation, collation and exchange of information especially for managing dynamic data and information within a BIM environment.

Proposed solutions rely on the use of both open and proprietary format. Indeed, the importance of using IFC is recognised mainly for its use as exchange information format in the public sector. However, several barriers are still encountered in the adoption of IFC, mainly due to the highly redundant IFC schema, to inconsistencies in the different assumptions and interpretations in expressing information, to the ambiguous quality of IFC certification processes and to the difficulty of users in personalising import and export options of IFC formats in BIM authoring platforms.

Therefore, as obstacles can be encountered in IFC adoption, a method is also tested for improving the information exchange process, based on the connection of information between models and sensors and between native and IFC models using external databases through automated processes.

Within the framework, according to defined model uses, the information content needs to be defined so that each piece of information is linked to corresponding entities in the IFC schema mapping smart objects from real and individual devices to BIModels, when adopting an open-BIM approach. Hence, a first issue (Issue 1) is considered, for extending the IFC schema in the case it does not cover the content defined according to defined model uses. Once the desired entities and properties are in the IFC schema, solutions (e.g. BACS-IFC converters) are proposed for updating BIModels with data gathered through sensors (Issue 2). However, if information is not required to be directly modelled into the BIModel, solutions are proposed for enriching the BIM content connecting information to external databases (Issue 3), based on proprietary tools.

For developing the proposed solutions, a comparison is needed between different environments for managing information, different workflows for sharing information and different formats for exchanging information.

Indeed, for storing data gathered through sensors in BIModels, the IFC schema needs to be extended for managing selected property sets of objects (Fig. 7.1). The proposed process deals with the definition of properties to be modelled and associated to selected objects (on behalf of clients), the setting of values through sensors and the validation of data through rules (e.g. by users), generating warnings through automated rules when values exceed defined thresholds. These thresholds are defined according to normative references or regulations (e.g. for what concerns CO_2 concentration, comfort conditions, occupancy) or on results of energy simulations.

After analysing the IFC schema, two different solutions are developed and compared for enriching objects with properties not included in the IFC specifications. The first solution extends the IFC schema through non-proprietary solutions, even if advanced IT-skills are required to the users. Therefore, a second proprietary solution is proposed.

In the first solutions, the information exchange process is developed in a non-proprietary environment, as the IFC is both generated and validated in an open-source editor based on the use of programming languages. Particularly, Python-encoded

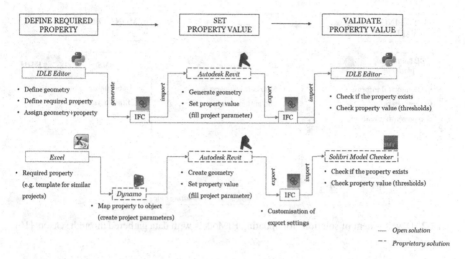

Fig. 7.1 Definition of information requirements according to defined model uses [16]

scripts are created and modified for creating objects and associated properties, according to the IFC schema. Once properties are defined, they can be filled within IDLE Editor or in the BIM authoring platform. Rules for validating the information content are developed as Python-encoded scripts. Those rules allow verifying if selected properties are correctly added to objects and if their values are set. Therefore, based on registered values, ad hoc rules generate warnings if properties are not included, if values are not filled or if values exceed defined thresholds.

As among the main barriers in the adoption of such an information exchange process there is the advanced IT-skills required to building operators for customising the scripts, a similar process is developed using proprietary solutions. In that case, information requirements are defined in an external database in the form of a list, where thresholds or limits to values are also fixed. In the list, properties are associated to the objects. Then, according to the used tools, through Visual Programming Environments, semi-automated processes enhance the import of such properties and the creation of defined property sets (included or not in the IFC Specification) in a BIM authoring platform. Then, connecting sensors to BIModels, the values of properties are automatically filled with registered data. In order to export created property sets (included or not in the IFC Specification) in an IFC file, customised options are set.

Comparing the proposed solutions, for adopting the non-proprietary solution, advanced IT-skills are required for generating and verifying the information requirements in Python. However, the creation and validation of properties and objects directly in the IFC ensures that all information is carried through the different stages.

Instead, the adoption of the proprietary solution is bounded by software constraints, even if it can be easily adopted and re-adopted in different projects and for different BIM uses, by modifying the list of properties and objects in the external database or spreadsheet.

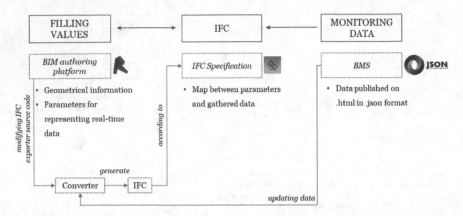

Fig. 7.2 Development of solutions for updating BIModels with data gathered through sensors [16]

Once the IFC schema has been extended so that selected IFC entities are adequately modelled in BIModels, data gathered through sensors need to be connected to these IFC entities (Fig. 7.2). Indeed, data gathered from sensors are generally stored into systems, as the BMS. By connecting BIM and BMS, both geometric and non-geometric information regarding the building (e.g. temperature and relative humidity) enrich the BIModels.

In order to exchange data between the BMS and the BIModel, a suitable converter to the IFC format has been written in Python, accessing the data into the BMS through its web services and outputting the IFC format.

In order to correctly map built-in parameters and the relative IFC objects (e.g. IFC sensor entities) and in order to avoid that missing information is registered in the exchanged model with respect to the original one, the BIM-IFC converter is customised, leveraging on the code developed by Autodesk, for adding specific properties to a BIM database in relation to information collected through sensors.

Hence, parameters created in the BIM authoring platform (that describe buildings and their properties in BIModels) are mapped to respective IFC properties.

Once those properties are correctly mapped and modelled, sensor-related information is stored in the underlined database that drives the BIModel and is exported in IFC.

Hence, sensor-related information is stored in the BIModel. An updating of those values, using data collected on the field, allows visualising and analysing them in suitable synoptic schemas.

Moreover, as it is possible to write information directly in an existing IFC file, an approach is proposed, considering the possibility to export a BIModel in IFC from a native format and then enrich it with the needed information through an automated process of information transcription (Fig. 7.3).

When users create a BIModel in a specific BIM authoring platform, they can provide different information in fields already defined in the tool or they can define additional fields for extending information content. Once the fields of the database are

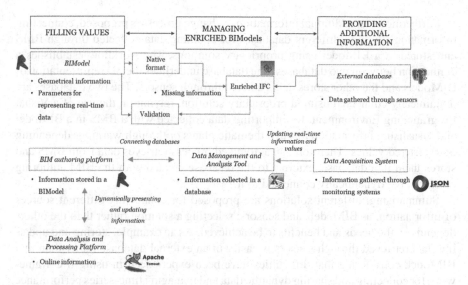

Fig. 7.3 Enrichment of BIM content connecting information to external databases [14]

filled with the required information, users export BIModels, translating the native model in IFC. That phase can be defined directly by the user without acting on the export-settings of commercially available software. However, once the IFC file is created using standard settings, problems because of missing information are encountered in the exchange process. Therefore, an automated process is developed in a Python-encoded environment for validation.

Indeed, the validation of information requirements is performed through customised scripts for reading the information of all the objects both in the native BIModel and in the IFC model, comparing the information in the two files and writing the missing information (if detected) in the structured database.

Once the set of information is written in the database, information can be stored in the database, linked to the model thanks to the unique IDs created by the BIM authoring platform for each modelled object. Otherwise, information can be maintained in the model itself, running a script able to read the information contained in the database and write the information directly in the IFC file.

Adopting the proposed automated process, information is not missing in the roundtrip between native format and IFC format.

Hence, the proposed solution is based on the complementarity of information in different databases. In fact, instead of dynamically linking the same information in two databases so that an information in a database updates its digital twin in the other database, a part of information is stored in a relational database and another part is stored in the BIModel. When information has to be exchanged, the two parts of information are merged.

The proposed solution can also be applied for controlling if values of selected properties are filled or if they exceed selected thresholds.

Furthermore, an additional information exchange process is proposed, connecting information stored in different databases for updating data collected through BMS and stored in a BIModel using proprietary solutions. The workflow establishes a connection between two databases; one database underlines the 3D representation of BIModels and the other stores data gathered through sensors. The two databases are dynamically connected using a proprietary solution, relying on the use of a Visual Programming Environment, for importing data collected with a BMS in a BIModel and visualising information through thematic plants or through warnings depending on registered values. Therefore, an automated process retrieves data from BMS and stores them in a database, external to the database created with the BIM authoring platform, but dynamically connected to it.

Summarising, different solutions are proposed for connecting different sources of information, as BIModels and sensors; selecting a solution rather than the others depends on the needs and benefits to be achieved. As an example, storing and updating data retrieved through sensors is easier in an external database rather than in a BIModel, considering that difficulties have been experienced in using IFC framework for collecting and sharing dynamic data and managing time-series performance data and considering the limited size of BIModels, as part of information is stored in an external database, dynamically linked to the BIModel. As associations are created between data included in a model and parameters based on external databases. Consequently, by changing a value derived from the input source, it is possible to dynamically change also the value in the BIModel. Hence, a real-time coordination is established between as-designed virtual models and as-delivered physical buildings.

The reproducibility and reusability of developed solutions have also been considered.

Moreover, as open formats are often required for exchanging information, nonproprietary solutions are provided, also testing the possibility to store information not considered in the IFC schema, presenting procedures for adding specific property sets to IFC. In this case, an advance IFC knowledge and skills for customising tools through programming languages have to be considered.

Furthermore, different IT-skills are required to building operators for exchanging information, outlining the need for establishing collaboration among AEC and IT operators and integrating AEC and IT skills for a digitalised built environment, e.g. in relation to the required ability to handle and interpret large datasets.

It is worth to mention that considerations are strictly related to the adopted tools for testing the information exchange processes.

Solutions for improving the information exchange process through different environments have been proposed for developing prototypes for accessing real-time information for different uses.

A just-in-time approach has been adopted for accessing and delivering building information in operation. Indeed, besides enriching BIModels to be modified and queried within BIM authoring platforms, the research proposes solutions also for making information stored within BIModels accessible without necessarily requiring the use of BIM authoring platforms. Particularly, solutions for accessing information on building conditions and user influence on them are proposed through web platforms and apps. Indeed, being the model and the information stored within it published on the web, users can always access the information so to be informed about monitored conditions and provided with control strategies.

7.7   Results

The chapter presents the role and value of data and information in tackling challenges within a digital and digitalised built environment from the perspective of owners and end-users.

Particularly, the research presents a framework for the engagement of users in building operation, based on the interaction between users and buildings in a bidirectional way. Indeed, awareness derived from an improved use of information (i.e. with reference to energy-hungry behaviours) through the integration of ICT solutions and BIM-based processes improves the on-going management and performance of buildings.

Particularly, real-time data collected during operational stages of buildings can be used for analysing the influence of user behaviours on building conditions (e.g. with reference to windows opening behaviours and setting of HVAC operations), simulating the effects of behaviours in different scenarios and proposing control strategies, mainly for reducing consumptions in building operation.

The real-time management of data and information ensures the establishment of a feedback loop among clients and users. Indeed, a link among clients and users aligns the supply chain with the demand side, resulting in a long-term building performance. The building process and the final product are not improved if clients equip buildings with sensors that do not provide useful and meaningful measurements to users. If sensors are wrongly located, the gathered data do not depict the real situation of rooms. If data are collected by devices with different time ranges, a complete picture of the building at a specific time is difficult to be provided; collecting data from different devices with the same timestamps for different indicators enhances their comparisons and correlations. Moreover, if clients collect data with a tight time range, users can access a near-instantaneous behaviour of the building, even if challenges can be encountered when collecting redundant data, catching and showing changes over time while minimising the dataset size. Redundant data are useless and slow down the comprehension of building behaviours.

Within that context, actors have a great role in defining rules and controlling information.

However, besides the created value, there are also challenges concerning data-related changes in building processes. Among those challenges, extra training, resources and time are required for managing great amounts of data. New tools and techniques improve the fruition of data, requiring at the same time new skills (e.g. in fields related to data mining, cognitive computing, programming languages) for building operators. Therefore, the building sector needs a reconfiguration also in term of required skills and profiles to building operators, shifting towards workers who are able to make sense of data [10]. Indeed, design firms generally lack of knowledgeable data experts who can intelligibly curate diverse data sources and tools according to the project needs [5]. Moreover, the reconfiguration is also related to the number of disciplines and stakeholders involved in the AEC sector for efficiently addressing, managing and integrating data across those disciplines [20]. Indeed, a multidisciplinary approach is required for considering occupant behaviours in buildings, crossing social and behaviour science, building science, sensing and control technologies, computing science, and data science [9].

The last challenge is due to contractual complexity and the uncertainty around who owns the data and the liability for the project outcome [20].

Within that context, the research highlights how the concept of building is evolving from being a financial product to becoming a service provider to support the needs of users, in relation to personal habits and individualised requirements. The chapter presents solutions to establish an interaction between buildings and their users. The interaction mainly derives by user awareness, achieved by providing meaningful information to users in the form of control strategies. Furthermore, the interaction is ensured through the connection between information collected through different systems within a building and with the surrounding environment, also integrating BIM processes and IoT paradigms.

Considering that BIM Level 3 is related to the development of new business models for asset design, delivery, operation and adaptation based on a wider use of service performance data, the research starts bringing together service-orientated and outcomes-based industry, through data from operations analysing and creating the learning feedback loops that industry needs to be able to deliver sustainable long-term improvements in asset performance. Main results achieved in the real-time management and information exchange processes could be the basis for the development of business models for manipulating open data, innovation and complex asset networks, creating an asset user/asset manager industry. Skills within the industry should focus on the flow and process of information procurement and transactions throughout the supply chain.

Indeed, by establishing a connection between as-designed virtual models and as-delivered physical assets, building processes rely on both performance-oriented design and validated-operations, where building behaviours and user activities are monitored, predicted performances and real measurements are correlated and, consequently, the building performance gap is estimated.

References

1. Becerik-Gerber B, Jazizadeh F, Li N, Calis G (2011) Application areas and data requirements for BIM-enabled facilities management. J Constr Eng Manage 138(3):431–442
2. Chen W, Chen K, Cheng JC, Wang Q, Gan VJ (2018) BIM-based framework for automatic scheduling of facility maintenance work orders. Autom Constr 91:15–30

3. Cavka HB, Staub-French S, Poirier EA (2017) Developing owner information requirements for BIM-enabled project delivery and asset management. Autom Constr 83:169–183. https://doi.org/10.1016/j.autcon.2017.08.006
4. Dave B, Buda A, Nurminen A, Främling K (2018) A framework for integrating BIM and IoT through open standards. Autom Constr 95:35–45. https://doi.org/10.1016/j.autcon.2018.07.022
5. Deutsch R (2015) Data-driven design and construction: 25 strategies for capturing, analyzing and applying building data. Wiley, Hoboken, NJ
6. Gallaher MP, O'Conor AC, Dettbarn JL, Gilday LT (2004) Cost analysis of inadequate interoperability in the U.S. Capital Facilities Industry, National Institute of Standards & Technology, Gaithersburg
7. Gerrish T, Ruikar K, Cook M, Johnson M, Phillip M, Lowry C (2017) BIM application to building energy performance visualisation and management: challenges and potential. Energy Build 144:218–228
8. Gruber TR (1995) Toward principles for the design of ontologies used for knowledge sharing? Int J Hum Comput Stud 43(5–6):907–928
9. Hong T, Taylor-Lange SC, D'Oca S, Yan D, Corgnati SP (2016) Advances in research and applications of energy-related occupant behavior in buildings. Energy Build 116:694–702. https://doi.org/10.1016/j.enbuild.2015.11.052
10. Kaur H, Lechman E, Marszk A (eds) (2017) Catalyzing development through ICT adoption: the developing world experience. Springer, New York
11. Kim K, Kim H, Kim W, Kim C, Kim J, Yu J (2018) Integration of ifc objects and facility management work information using semantic web. Automation in Construction 87:173–187. https://doi.org/10.1016/j.autcon.2017.12.019
12. Kiviniemi A (2013) Value of BIM in FM/OM—why have we failed in attracting owners and operators? https://wiki.aalto.fi/download/attachments/80938503/BIM%20in%20FM%26OM%20Kiviniemi%2020130404.pdf?version=1&modificationDate=1372406821000&api=v2. Last accessed on 22 July 2019
13. McArthur JJ, Shahbazi N, Fok R, Raghubar C, Bortoluzzi B, An A (2018) Machine learning and BIM visualization for maintenance issue classification and enhanced data collection. Adv Eng Inform 38:101–112. https://doi.org/10.1016/j.aei.2018.06.007
14. Mirarchi C, Pasini D, Pavan A, Daniotti B (2017) Automated IFC-based processes in the construction sector: a method for improving the information flow. In: LC3 conference, Heraklion
15. Pasini D (2018) Connecting BIM and IoT for engaging users in building operation. J Struct Integrity Maintenance
16. Pasini D, Mastrolembo Ventura S, Bolpagni M (2017) BIM-based process for managing property sets of objects extending the IFC schema. In: ISTeA conference, Firenze
17. Pasini D, Reda F, Häkkinen T (2017) User engaging practices for energy saving in buildings: critical review and new enhanced procedure. Energy Build 148:74–88. https://doi.org/10.1016/j.enbuild.2017.05.010
18. Peng Y, Lin JR, Zhang JP, Hu ZZ (2017) A hybrid data mining approach on BIM-based building operation and maintenance. Build Environ 126:483–495. https://doi.org/10.1016/j.buildenv.2017.09.030
19. Pishdad-Bozorgi P, Gao X, Eastman C, Self AP (2018) Planning and developing facility management-enabled building information model (FM-enabled BIM). Autom Constr 87:22–38. https://doi.org/10.1016/j.autcon.2017.12.004
20. Qabshoqa M, Kocaturk T, Kiviniemi A (2017) A value-driven perspective to understand Data-driven futures in Architecture. eCAADe
21. Sattenini A, Azhar S, Thuston J (2011) Preparing a building information model for facility maintenance and management. In: 28th International symposium on automation and robotics in construction, Seoul, South Korea, pp 144–149
22. Young NW, Jones SA, Bernstein HM (2007) Interoperability in the construction industry

Printed in the United States
By Bookmasters